Vintage Brooklyn Auto Ads Vol:4

By Robert A Henriksen

Copyright © - 2018 by Robert A Henriksen

All rights reserved.

If you are into vintage cars, this is the book for you! If you are into vintage car advertising, this is definitely the book for you. The Vintage Car Show releases a 4-volume series, **Vintage Brooklyn Auto Ads**. These ads are from the "heydays" of Brooklyn and the automotive industry. The books are a mother load of vintage automotive advertisements. For more automotive history visit the website thevintagecarshow.com

Vintage Brooklyn Auto Ads have over 400 different ads in each book, from the early days of the automobile up to the 50s, with dealership ads, used car ads, oil ads, gasoline ads, tire ads, etc. It contains some clever and creative advertising, as well.

These books are part of the Brooklyn Book Series, which also includes neighborhood books, **The Park Slopian**, **Growing Up in Bay Ridge**, **Growing Up Brooklyn**, and a 3-volume set (**Made in Brooklyn and Other Ephemeral**) More titles will be coming. Check out the website for future updates at brooklynpast.com.

Coming soon is a 14-volume set of **Brooklyn vintage Ads** from the 20th century. This set features advertising from some of the most prestigious companies of the day. See if you can spot a few who still produce products 100 years later.

Vintage Brooklyn Ads Vol #2 is a collection and selection of over 400 vintage ads, matchbooks, and other types of Brooklyn Ephemeral dating back to the late 1800's..
Made is Brooklyn Ads Vol #1. 2 great books over a thousand vintage ads.

Brooklyn the series: In which neighborhood in Brooklyn did you grow up? Do you yearn for the good old days? This series of neighborhood books is the closest you'll get to reliving the past—or, if you're not from Brooklyn, experiencing the fun, the excitement and the tragedies, right from the memories of the kids and the parents who lived there during the fascinating middle of the twentieth century.

These books walk you through the various neighborhoods in classic 1950s through 1980s Brooklyn, detailing the iconic things of our time. From the doctors who delivered us to the schools from which we graduated, from the playgrounds and parks in which we played to the street games we made up by ourselves, from all the great toys we had that have since been replaced by sharper technology, our first bikes to our first cars—we remember it all.

Take a trip down memory lane with these coffee table books, written in an enjoyable, accessible, social media style. Revisit all the best places we ate; remember all the silly slang and the nonsensical stuff we used to say. As with every passing generation, we cling to the things that defined our youth. However, we who grew up in the fifties through the eighties experienced some of the most timeless pop culture in history, and this extraordinary series will allow you to share that with your children using a language they understand: social media.

This wonderful series of coffee table books contrasts bonding over modern social media with longing for the past. If aliens came down to Earth sometime in the future and found these books, they would act as a time capsule—a treasure trove of memories of mid-twentieth century pop culture, and a demonstration of twenty-first century netspeak and social media usage to boot.

Use these books to share your memories with friends who grew up elsewhere, and as you pore over the pages, find all the similarities and differences in the toys you owned versus the mom and pop businesses that you frequented. Designed as coffee table books, they can be picked up and put down at your leisure; they do not need be read cover to cover. If you own these books and display them in your living room, your friends will want to visit you more often! Find them all at **www.brooklynpast.com**

   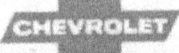

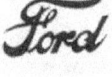

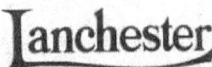 

WHY THE HAYNES IS FIRST

The Haynes Factory was the first automobile factory to be built in America (1893).
The Haynes was the first to adopt low-tension make-and-break ignition (1895).
The Haynes was the first to use nickel steel and aluminum alloy in a car.
The Haynes was the first to adopt side entrance bodies and large wheels.
The Haynes IS first to adopt the roller pinion and bevelled sprocket direct drive, making possible the combination of shaft drive and high power.
The Model T Chassis on the car shown herewith is the same chassis that proved its merit in the Vanderbilt Cup Race, where against foreign and American cars of twice its H. P., it made such a wonderful showing for speed and endurance.

The Haynes Standard 50 H. P. Touring Car for 1907, Model "T," the highest-powered shaft-driven car built. Price $3,500.
35 H. P. Touring Car, $2,500.

Send at once for full information and specifications. Address Desk F-13.

HAYNES AUTOMOBILE CO., KOKOMO, IND.

Oldest Automobile Manufacturers in America.   Members A. L. A. M.

New York—1715 Broadway   Chicago—1420 Michigan Avenue

AN ANNOUNCEMENT THAT'S WORTH WHILE

The Gearless Transmission Company

Will exhibit for the first time, at **The Boston Auto Show**, the following cars, comprising their 1907 line. A glance at the specifications will convince you that they will be **the sensation of the year**, combining as they do, the most advanced construction in every detail. See them, Space 221, Dept. C., Mechanics Hall.

Model 60 — Sixty Horse Power

4 Cylinders, 2 Cycle, 5-in. x 5-in., Water Cooled.
The Gearless, Valveless Car
Wonderful Power
Simplest Control
Beautiful in Design
Superb Finish

List, $3,250

Model 75

Seventy-five Horse Power
6 Cylinders, 4 Cycle, 4¾-in. x 5¼-in., Water Cooled.
"THE BIGGEST SIX OF THEM ALL."
Seats Seven Passengers.
Straight Line Body.

List, $4,000

Send for Advance Catalogue. The greatest proposition on wheels. Both Models Equipped with the 1907 Model Gearless Transmission, "The Direct Drive Friction Transmission."

Gearless Transmission Co. : Rochester, N. Y. : Motor Car Dept.

CONDUITE INTÉRIEURE 4 PLACES SUR CHASSIS **6** CV

Constant Performance Everywhere

What scientific investigation and engineering skill have dictated in Remy equipment, practical experience has borne out to the fullest degree.

Remy has the advantage of nineteen years' steady, progressive growth, side-by-side with the automobile itself, and last year manufactured over two hundred thousand Remy products, to say nothing of more than a million that went into service before.

Yet even with a successful record like this, Remy engineers are still at work seeking to raise still higher the standard of "Products of Constant Performance," in order to anticipate future requirements.

It is this record and this constant progress that inspire confidence when the salesman answers your Starting, Lighting and Ignition question by saying, "Remy."

STARTING LIGHTING IGNITION SYSTEMS

REMY ELECTRIC COMPANY
Motor Equipment Division,
Detroit, Michigan

General Offices and Factories:
Anderson, Indiana

Laboratories:
Detroit, Michigan

Tractor Equipment Division, Chicago, Illinois

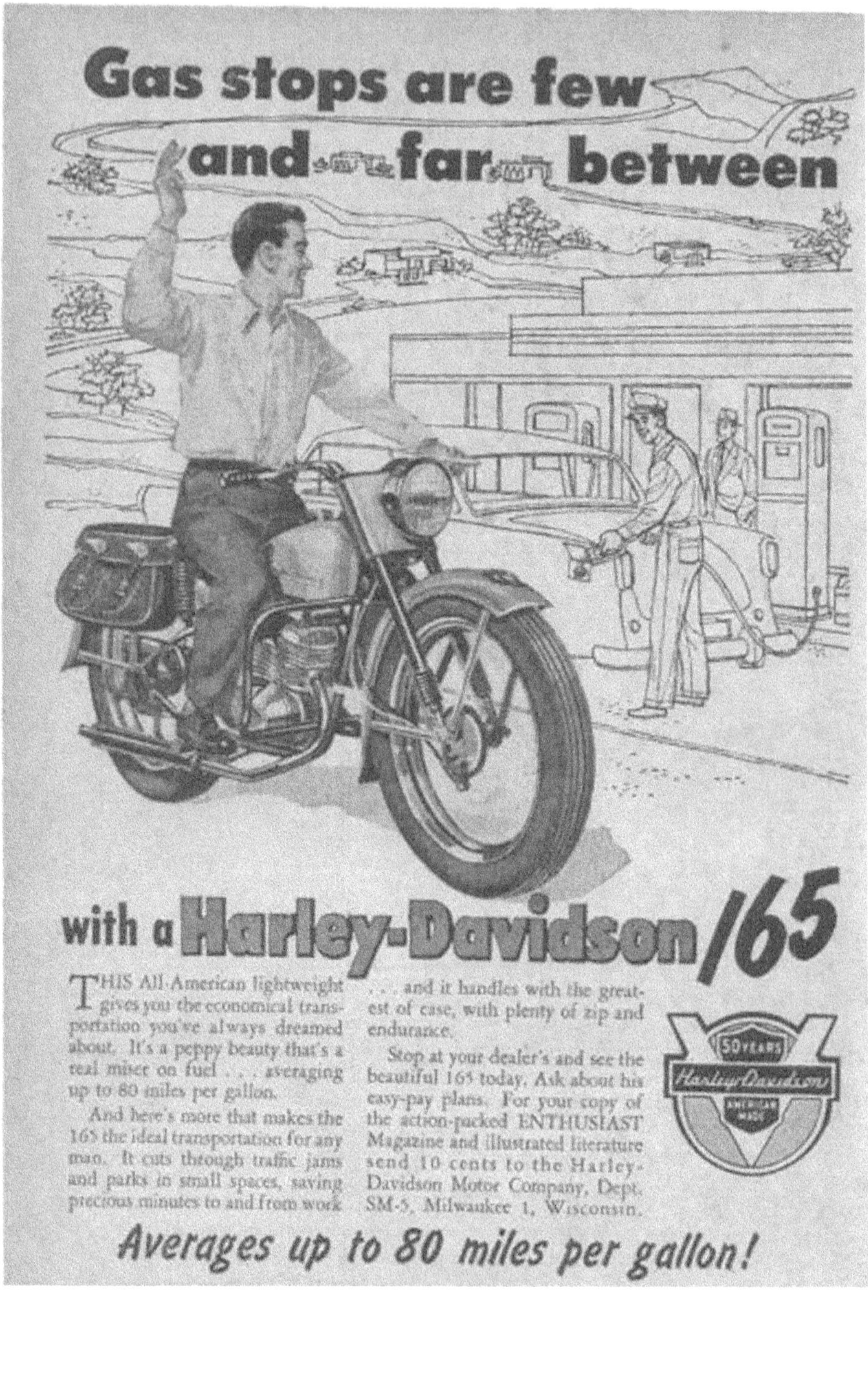

Gas stops are few and far between

with a Harley-Davidson 165

THIS All-American lightweight gives you the economical transportation you've always dreamed about. It's a peppy beauty that's a real miser on fuel... averaging up to 80 miles per gallon.

And here's more that makes the 165 the ideal transportation for any man. It cuts through traffic jams and parks in small spaces, saving precious minutes to and from work ... and it handles with the greatest of ease, with plenty of zip and endurance.

Stop at your dealer's and see the beautiful 165 today. Ask about his easy-pay plans. For your copy of the action-packed ENTHUSIAST Magazine and illustrated literature send 10 cents to the Harley-Davidson Motor Company, Dept. SM-5, Milwaukee 1, Wisconsin.

Averages up to 80 miles per gallon!

HAYNES 1921 CLOSED CARS

Utmost in beauty, luxury and utility — $1,000 underpriced

NOW, when the buyer at last is asking: "What am I getting for what I pay?" the advantage of the Haynes selling policy becomes increasingly evident. Enthusiastic Haynes owners have told us all along that the Haynes is $1,000 underpriced. The Haynes principle of building for the future has held good. We have been and are satisfied to produce the choicest car of its class and sell it at a price that is fair to the buyer and to ourselves.

The seven-passenger Haynes Suburban and the five-passenger Haynes Brougham richly deserve the high praise accorded them. Among closed cars they establish a class of their own. Quietly rich in finish and fittings, as such cars should be, they are distinguished in line and completely desirable in appearance. They are far and away beyond anything to be expected in their price-class.

A detailed description of the many superiorities of construction and design and of the thoughtful conveniences installed in each car is obviously impossible here. A personal inspection of these closed cars is invited and urged. To secure prompt delivery an immediate reservation is recommended.

THE HAYNES AUTOMOBILE COMPANY
KOKOMO, INDIANA   U.S.A.
Export Office: 1915 Broadway, New York City, U.S.A.

HAYNES
CHARACTER CARS
Beauty — Strength — Power — Comfort

1893 ✦ THE HAYNES IS AMERICA'S FIRST CAR ✦ 1920

Starting Military Vehicles on the Road to Victory

PRODUCT OF DELCO-REMY

Starting Essential Cars on Their Wartime Jobs

The 3 Fundamentals of BATTERY CARE
1 Add water regularly
2 Keep connections clean and tight
3 Recharge when necessary

Delco batteries are today meeting the needs of the armed forces—in planes, tanks, combat cars, trucks and other military vehicles.

Millions more are helping to maintain essential transportation at home—in cars, trucks and buses.

This double responsibility means that every Delco battery built today has an important assignment to carry out on the fighting front or the home front. It gives every civilian car owner this urgent command:

Don't waste battery life by neglect . . . don't squander it by buying a new battery when your present one is still dependable.

See your Delco battery dealer every two weeks for inspection and service. His good judgment will tell him when your present battery has outlived its usefulness. Your good judgment will tell you to replace with a Delco battery . . . for extra starting power, for long life, for dependability.

When You Must Replace REPLACE WITH A DELCO BATTERY

Delco batteries are built for every make and model automobile, as well as for trucks, buses and tractors. They are sold by 40,000 dealers under the direction of United Motors Service.

★ DON'T LET UP, OR YOU'LL LET A FIGHTER DOWN—KEEP ON BUYING BONDS ★

DELCO-REMY ★ WHEREVER WHEELS TURN OR PROPELLERS SPIN

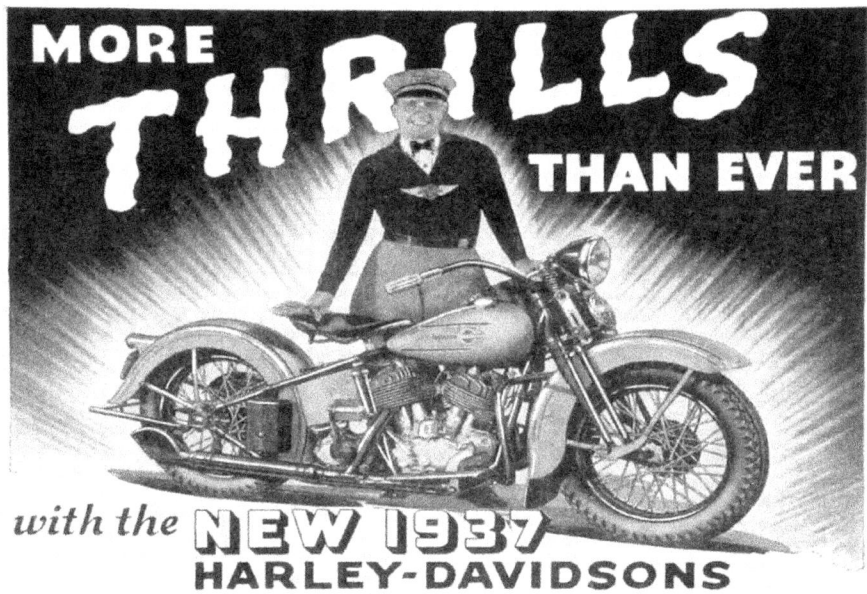

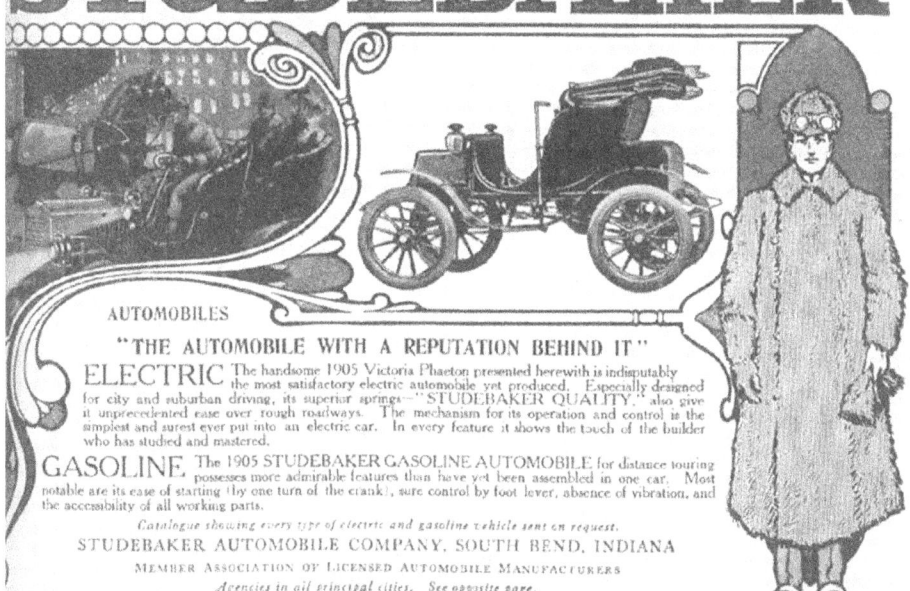

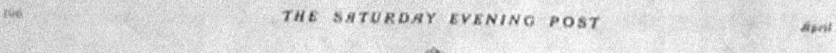

Salient STEPHENS *Six*

THE longer our owners drive the Stephens Salient Six the greater grows their appreciation of its extraordinary economy and dependability.

They measure its service in scores of thousands of miles. Their enthusiasm is reflected in the admiration owners of other cars so frankly express for the Stephens.

The owner's pride is the result of the car's abundant resources in power, beauty, economy and reliability. Our owners are never content until they have sold the car's virtues to their friends.

This unusual tribute to the automobile craftsmanship of the Stephens Works has been won and held only through the car's embodiment of that conservative distinction which marks all things of permanent value.

Stephens Motor Works ·· Freeport, Illinois

for Economical Transportation

Economy

Low in price—low in cost of operation —with service available everywhere, Chevrolet is recognized as the foremost car "for Economical Transportation."

CHEVROLET MOTOR COMPANY, DETROIT, MICHIGAN
DIVISION OF GENERAL MOTORS CORPORATION

Touring - $525
Roadster - $525
Coupe - - $715
Coach - - $735
Sedan - - $825
Commercial Chassis $425
Express Truck Chassis $550

ALL PRICES F. O. B.
FLINT, MICH.

QUALITY AT LOW COST

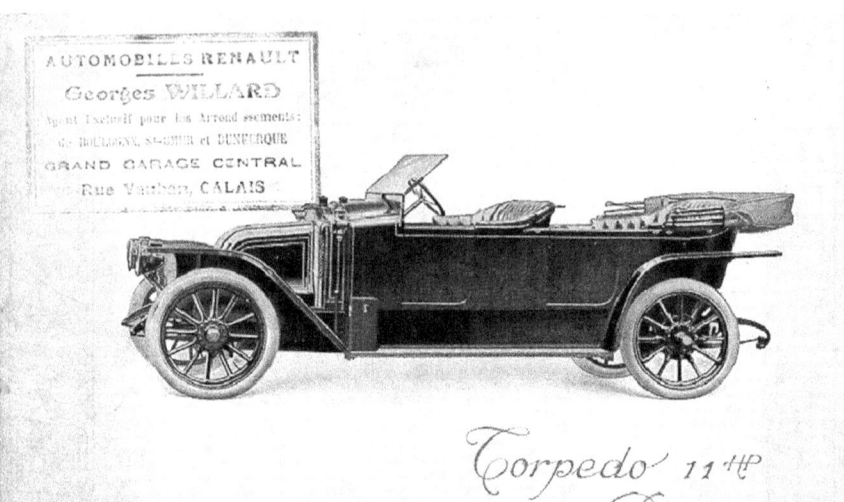

Torpedo 11ᵗᴴ Renault

The MG Sports

THE STILES THREESOME.

TWO SEATER AND DICKEY SEAT. CONCEALED HOOD. SUITABLE FOR MOUNTING ON 12/70 M.G. MAGNA AND D2 TYPE MIDGET CHASSIS.

STILES LTD.,   3, BAKER STREET, W.1.

A car of such manifest and extraordinary excellence—a chassis so costly—that it will upset all your previous notions of which is really the finest car made in America. We urge upon you nothing but this—**ride in the Craig-Toledo.** We will abide by the results.

THE CRAIG=TOLEDO MOTOR COMPANY
TOLEDO, OHIO

How SINCLAIR helps NEW CAR OWNERS *"Beat the Devil"*

Owners of new cars, and older automobiles too—36,000,000 are on the road in 1951—are finding out that one of their deadliest enemies is *rust*. This red devil, which gets into gasoline station fuel pumps as well as the entire fuel system of your car when moist air in the fuel tank condenses, does costly damage to the tank, carburetor, fuel pump and other vital car parts.

Sinclair found the way to "beat the devil" when it developed RD-119, the amazing new gasoline ingredient that covers metal surfaces with an invisible *rust-proof* coating. An *exclusive* Sinclair discovery, RD-119 provides superlative anti-rust protection when added to gasoline and other petroleum fuels.

RD-119® is now standard in Sinclair gasoline—to protect both new and old cars from power-stealing rust. It is another outstanding result of Sinclair's progressive research and another in the growing list of reasons why Sinclair is . . . "a great name in oil."

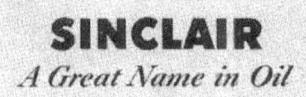

SINCLAIR
A Great Name in Oil

SINCLAIR OIL CORPORATION · 630 FIFTH AVENUE · NEW YORK 20, N.Y.

Mention the National Geographic—It identifies you

Salient
STEPHENS
Six

THE longer our owners drive the Stephens Salient Six the greater grows their appreciation of its extraordinary economy and dependability.

They measure its service in scores of thousands of miles. Their enthusiasm is reflected in the admiration owners of other cars so frankly express for the Stephens.

The owner's pride is the result of the car's abundant resources in power, beauty, economy and reliability. Our owners are never content until they have told the car's virtues to their friends.

This unusual tribute to the automobile craftsmanship of the Stephens Works has been won and held only through the car's embodiment of that conservative distinction which marks all things of permanent value.

Stephens Motor Works ·· Freeport, Illinois

for Economical Transportation

Economy

Low in price—low in cost of operation—with service available everywhere, Chevrolet is recognized as the foremost car "for Economical Transportation."

CHEVROLET MOTOR COMPANY, DETROIT, MICHIGAN
DIVISION OF GENERAL MOTORS CORPORATION

Touring	$525
Roadster	$525
Coupe	$715
Coach	$735
Sedan	$825
Commercial Chassis	$425
Express Truck Chassis	$550

ALL PRICES F. O. B.
FLINT, MICH.

QUALITY AT LOW COST

THE
Price, $4,500, Complete

ACME

SEXTUPLET

is the **ACME** of **RELIABILITY, SPEED, STYLE,** and **COMFORT,** with cost of maintenance extremely low. Write or Telephone **1454** Flatbush for demonstration at any time.

J. W. MEARS
7-9 Ocean Parkway.

GARAGE with best equipment and skilled mechanics

The Stearns
BEST OF THE WORLD

EXAMINE THE RECORDS

FORT GEORGE HILL CLIMB

Best Times Made by Any Stock 4-Cylinder Cars.			
STEARNS "30" 1st	- - - - -	42 1-5	$4,600
STEARNS "30" 2d	The Allen-Swan Co.	42 2-5	$4,600
STEARNS "30" 3d	The Allen-Swan Co.	42 2-5	$4,600
STEARNS "30" 4th	The Allen-Swan Co.	42 4-5	$4,600
SIMPLEX "50"	- - - - -	43 3-5	$5,750
SIMPLEX "50"	- - - - -	44	$5,750
RENAULT "45" Special Racing Runabout		49	$8,000

THE ALLEN-SWAN CO.
1287-1291 BEDFORD AVENUE
BROOKLYN
Telephone: 4192 Bedford

LEW H. ALLEN, PRESIDENT AND GEN'L. MANAGER
HALSTEAD SWAN, SECRETARY AND TREASURER
HOWARD DRAKELEY, SALES DEPARTMENT

Model "LC" 14 H. P. $825.

"The Dollar for Dollar Car"

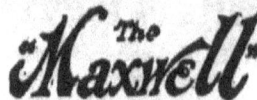

IS THE CONNECTING LINK
BETWEEN MOTORING
AND
SATISFACTION

UNFAILING RELIABILITY

I. C. KIRKHAM
Exclusive Distributer for Long Island
1060 Bedford Ave., cor. Clifton Pl., Brooklyn
Phone: 4300 Bedford

ELLENBECK & MULLER
AUTOMOBILE GARAGE
912 Bedford Ave.   **Telephone 4647 Wmbg.**
BROOKLYN
Agents for the

SIX MODELS
35 h. p. $2,750   TO   96 h. P. $7,500

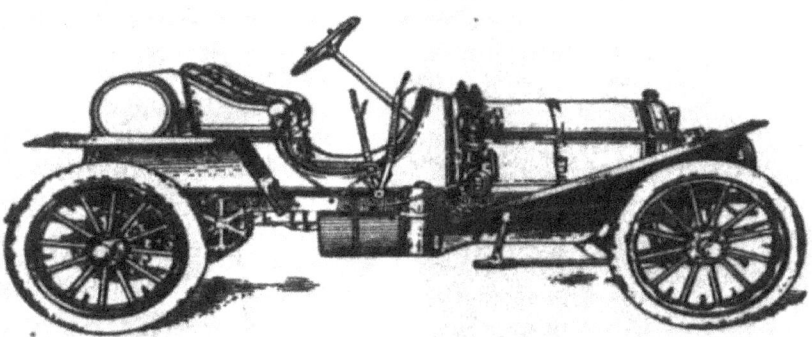

THE APPERSON JACKRABBIT
The Fastest Car for its Power in the World

LATE APPERSON VICTORIES

March 6—Altadena Hill-Climb—World's record 1 2-5 miles in 1.35 up; 11 4-5% grade; beating nearest competitor 19 seconds.

March 18—Savannah—Little 30 h. p,. $2,750 model wins 180 mile race in 3h. 35m. 41sec. Average speed over 50 miles an hour, beating nearest car over 18 miles.

March 19—Savannah—Jackrabbit model, from hopeless fifth position after breaking oil pipes and having tire trouble finished strong in second position.

A LONG DEMONSTRATION WILL CONVINCE YOU.

THE INCOMPARABLE
WHITE
THE CAR FOR SERVICE

A TOURING CAR IN TRUE SENSE OF THE TERM

The man who gets the most pleasure from his touring car is not the man who limits his touring to the macadam roads; for the most interesting sections of the country and those of the greatest natural beauty lie, for the most part, beyond the regions of improved highways. For that reason, there is no quality of a motor car more important than the ability to traverse bad roads.

In unique degree, the White possesses the qualities of a "bad roads" car. Owing to the perfect flexibility of the engine, the White tourist can accommodate the speed of his car, yard by yard, to the condition of the road, speeding up on each little stretch of good road, and slowing down for each hole and "thank-ye-ma'am"—without shifting of gears or any manipulation except of the throttle. The tremendous pulling power of the White engine under all conditions means immunity from getting stuck in the mud or sand. Running through deep water, as in fording streams, is easy for a White. And as for climbing grades in mountainous regions—there is no other machine which can approach the White in hill-climbing qualities.

Drive a White Steamer and see the country

THE WHITE COMPANY
CLEVELAND, OHIO.

NEW YORK CITY, Broadway at 62d Street
SAN FRANCISCO, 1460 Market Street
PHILADELPHIA, 629-33 North Broad Street
BOSTON, 320 Newbury Street
CHICAGO, 240 Michigan Avenue
CLEVELAND, 407 Rockwell Avenue
PITTSBURG, 138-148 Beatty Street

BIG SIX
Wins Fort George Hill Climb, N. Y.

Time, 37 3/5 seconds

The Big Six Stevens-Duryea made the best time of the day for regular stock cars.

LIGHT SIX

Repeats its victory of 1907, winning event F.

I. M. ALLEN CO.
116 So. Portland Ave., Brooklyn, N. Y.
Tel. 4026 Prospect

Manufactured by Stevens-Duryea Company Chicopee Falls, Mass., U. S. A.

| NEWPORT Establishment Opens **June 1** Address 100 Bellevue Ave. | # BAKER ELECTRIC VEHICLES | SPECIAL EXHIBIT at Waldorf-Astoria Showrooms **April 20=25** INCLUSIVE |

Have stood the test of time.

All the new models are built by the Baker Motor Vehicle Company of Cleveland, with that thoroughness and care which is characteristic of this concern, in a plant which is the largest and best equipped of its kind in the world. Electric Vehicle manufacture is the sole and single purpose in this greatest of automobile plants and we stand ready to demonstrate that Baker Electric Vehicles have no superior in this or any other country.

Immediate Deliveries

ALSO
Landaulets, Broughams, Coupes, Victorias, Stanhopes, Suburbans, Depot Carriages, Imperials, etc.

BAKER ROADSTER
A smart, novel and serviceable all around car for pleasure, riding or business use; strong, speedy, powerful. It has a greater mileage in any of its various speeds than any other electric car can demonstrate with an equipment of regular stock batteries and tires.

BAKER MOTOR VEHICLE COMPANY of N. Y.
1790 Broadway, cor. 58th St.

BAKER QUEEN VICTORIA
A miniature Victoria embodying all the graces of this type of carriage with the easy and quick action of the runabout, it lends distinction to the drive in the park, the social call, the shopping trip.

MERCANTILE MOTOR CAR CO. (Inc.)

Successors to Long Island Motor Car Co.

General Machinists

Export Repairing of all Cars

FIRST-CLASS CARS Fully Equipped for Renting

Telephone, 3538 Prospect.

Automobiles Stored

Repaired and sold on commission

Dealers in Second-hand Automobiles

368 Cumberland Street,   **Brooklyn, N. Y.**

PENNSYLVANIA

The way to "size up"

THERE ARE NO HILLS

The way to "size up" a proposition is to be "on the ground." If you will let us know where your "ground" is we will get a "Pennsylvania" on it—i. e., if you think of buying a really good car—advertising rhetoric don't make "the goods"—there is nothing so bad but what good *can be said* of it.

50 H. P.
114 W. B.

$2,800

Grant Square Auto Co.
Brooklyn and L. I. Distributers
1378 Bedford Avenue

TYPE E 1908

The Car of Durability that has Stood the Test of Time

Every necessary feature requisite to produce the Perfect Touring Car is found in the

Type "E" and Type "I" Models for 1908
DEMONSTRATIONS

The I. S. REMSON M'f'g Co.
Sole Agents for Brooklyn and Long Island
Garage—754-760 Bedford Avenue

JOS. D. ROURK
1001-3 Bedford Ave.
Brooklyn and L. I. Agt. Cadillac Cars
'Phone, 3730 Bedford   OPEN EVENINGS

19,000

CARS
NOW IN USE

Why?
Call and we will show you

PALMER-SINGER

All cars sold by us are patent and guaranteed   licensed under Selden for one year.

Palmer-Singer Six-Sixty Runabout 6 cyl., 60 H. P. $2,850

Palmer-Singer Town and Country Car, 28-30 H. P., $3,000

Palmer-Singer Skimabout, 28-30 H. P., $1,950

Palmer-Singer Six-Fifty Racing Car, 6 cyl., 50 H. P., $2,450

Palmer-Singer Four-Forty, Seven Passenger Touring Car, 40 H. P., $4,000

The Palmer-Singer Skimabout Is on Exhibition in Our Salesrooms. It is the Motor Car Sensation of the Year

The "Skimabout" is a town runabout that will make forty-five miles an hour over country roads, climb any climbable hill, outrun and outwear most runabouts and still give a town and traffic service that none of them can give. It is the trimmest, most stylish and most attractive medium-power runabout on the market. For the man who has little chance to go touring far afield it fills a long felt need.

Nickel steel throughout. Imported F. & S. ball bearings exclusively. Bosch high tension magnetos and multiple disk clutches in all models. Nickel steel, double and single drop frames. Drop forged I beam, nickel steel front axle —four-speed selective type, sliding gear transmissions with direct drive on third speed. All brakes equalized, all expanding type and on rear wheels. Universal joints on all steering connections. All types shaft driven, all moving parts inclosed in dustproof cases.

Sole distributors the Simplex   **PALMER & SINGER MFG. CO.**   **Metropolitan distributors the Selden**
1620-22-24 Broadway, N. Y.

The Car that Defends America Against the World

Thomas Town Cars

BRIARCLIFF TROPHY
Thomas Car a Contestant

Roadster   Touring Car

THOMAS 6 CYLINDER CARS
The Finest and Most Powerful Motor Cars Built
Write for Catalogue

HARRY S. HOUPT CO.
63d and Broadway, New York   213 Clinton Avenue, Newark, N. J.

No Other Car
Foreign or American
Is Better Than

The 1908
Stearns
Best of the World

Designed by Engineers
Built by Mechanics
Tested by Experts

Consequently Operated with
Satisfaction to All

LET US DEMONSTRATE

THE
ALLEN-SWAN
CO.

BEDFORD AVENUE—1287-1291
PHONE—BEDFORD 4192

BROOKLYN

LEW. H. ALLEN . . President and Gen'l Mgr.
HALSTEAD SWAN . . . . . Sec'y and Treas.
HOWARD DRAKELEY, Sales Dept.

ELLENBECK & MULLER
AUTOMOBILE GARAGE

912 Bedford Ave.   Telephone 4647 Wmbg.
BROOKLYN

Agents for the

SIX MODELS
35 h. p. $2,750   TO   96 h. P. $7,500

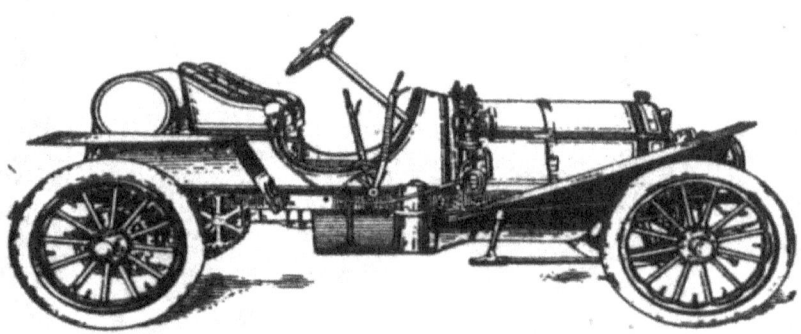

THE APPERSON JACKRABBIT
The Fastest Car for its Power in the World

LATE APPERSON VICTORIES

March 6—Altadena Hill-Climb — World's record 1 2-5 miles in 1.35 up; 11 4-5% grade; beating nearest competitor 19 seconds.

March 18—Savannah— L'ttle 30 h. p,. $2,750 model wins 180 mile race in 3h. 35m. 41sec. Average speed over 50 miles an hour, beating nearest car over 18 miles.

March 19—Savannah—Jackrabbit model, from hopeless fifth position after breaking oil pipes and having tire trouble finished strong in second position.

A LONG DEMONSTRATION WILL CONVINCE YOU.

Why the Ful-Floteing Seat Makes This Motorcycle Comfortable

THE ordinary motorcycle equipped with a rigid unadjustable leaf or coil spring that must be strong enough to support a 300 lb. man or break, makes mighty uncomfortable riding for a rider of lesser weight, just the same as a light auto equipped with the heavy, stiff springs used on auto trucks would make riding for its occupants any thing but comfortable. The adjustability of the springs in the FUL-FLOTEING SEAT, an exclusive patented feature of the

HARLEY-DAVIDSON

makes the Harley-Davidson the one motorcycle that is absolutely comfortable for every rider, regardless of weight. A comfort device to be worthy of its name, must not only protect the rider from the jars and jolts of rough roads, but must also overcome an equally objectionable feature, the rebound, so common to the ordinary motorcycle.

The FUL-FLOTEING SEAT of the Harley-Davidson virtually floats or carries the weight of the rider between two concealed compressed springs. One of these springs assimilates the jolts and jars due to bumps; the other absorbs the rebound. Thus the rider feels no jars, vibrations or rebound, but floats along in comfort.

The strength of these springs can be adjusted by a few turns of a tension nut so that they will operate perfectly under the weight of the rider. Thus the rider of a Harley-Davidson experiences pleasure and comfort going over the roughest road as he does when riding on the smoothest boulevard.

When you purchase a motorcycle—get a comfortable one with an adjustable FUL-FLOTEING SEAT—a Harley-Davidson. Ask the nearest Harley-Davidson dealer for a demonstration or write for catalog.

HARLEY-DAVIDSON MOTOR COMPANY
PRODUCERS OF HIGH GRADE MOTORCYCLES FOR OVER ELEVEN YEARS

331 B Street   MILWAUKEE, WISCONSIN

After a vehicle tire has persistently made good for over twelve years, it isn't necessary to do more than remind you of the name—

Kelly-Springfield

Made at Akron, Ohio. Sold by carriage manufacturers everywhere. "Rubber Tired" is a book about them. Sent free on request.

CONSOLIDATED RUBBER TIRE CO. New York Office, 20 Vesey St.

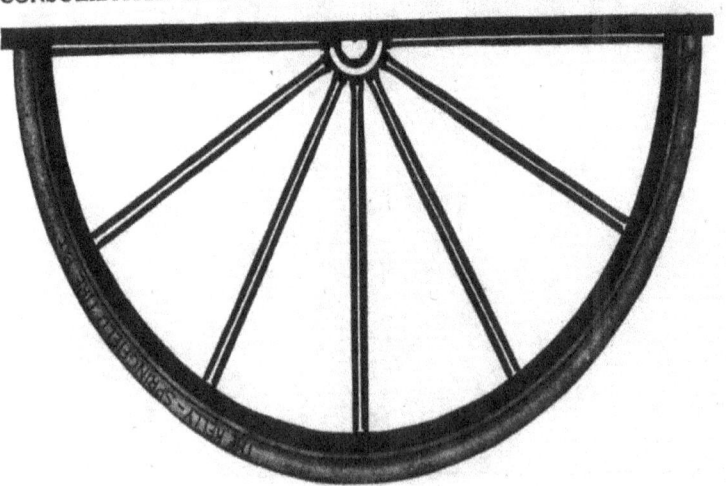

1911 Buick CLOSED CARS

We have sold more closed cars in the Metropolitan District this fall than any company engaged in the automobile business.

REASON No. 1

Our 1911 Model 41 Limousine has a four cylinder motor, cylinders cast in pairs, 4½ inch bore by 5 inch stroke and A.L.A.M. rating, 32.4 horse power. Wheel base, 116 inches, wheels 36 x 4 inches, and the body has five inside seats, all facing forward.

Other medium price, medium sized cars that give good service, have less style, finish and comfort and very much less power for touring, hill climbing, pulling through snow, and quick action in city traffic, and they cost more than $2,500.00.

REASON No. 2

Our large 1911 Model 7 Limousine has a four cylinder "T" type motor, cylinders cast in pairs, 5 inch bore by 5 inch stroke, and A.L.A.M. rating, 40 horse power. Wheel base 122 inches. Wheels 36 x 4 inches front, 36 x 4½ inches rear. The body has ample room for comfort and five inside seats.

These cars absolutely compete in size, material, workmanship, power, design, finish, comfort and A.L.A.M. horse power with $5,400 and $5,600 closed cars. The money you save by purchasing a Buick will pay the chauffeur, garage, tire, fuel and repair bill for a year. In the face of these concrete facts, can you, as a business proposition, purchase any other car?

Our extensive branch house organization, with shops, stock rooms and unequalled facilities, constitutes the safest guarantee of service offered by the industry.

BUICK MOTOR COMPANY,

NEW YORK
Broadway at 55th St.

BROOKLYN
42 Flatbush Ave.

NEWARK
222 Halsey St.

The "Maxwell" 1910

Model Q, 4 Cyl., 22 H.-P., Sliding Gear Transmission $850.00. Magneto Equipment. 5 Bodies to Choose From.

DEMONSTRATIONS BY APPOINTMENT.

I. C. KIRKHAM,

Tel. 4300 Bedford   1060 Bedford Ave.

The New O-te-sa-ga

on Otsego Lake

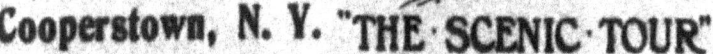

Cooperstown, N. Y. "THE SCENIC TOUR"

The beautiful region made famous by J. Fenimore Cooper, the Novelist. This new and perfectly appointed hotel will open for its first season July 12th, under the personal management of J. D. PRICE OF

ANDERSON & PRICE,

Hotel Ormond and Bretton Inn, Ormond Beach, Fla., and Bretton Woods Hotels, White Mountains, Hotel Bretton Hall and Hotel Seymour, New York City: Exceptional fishing, boating, golf, tennis and excellent automobile roads. ON "THE SCENIC TOUR."

For booklet, information and Automobile Road Maps, address New York Office, HOTEL BRETTON HALL, Broad'way and 86th Street.

BRIGHTON BEACH MOTORDROME

Friday and Saturday, Aug. 27-28

Rain or Shine

24 HOUR AUTOMOBILE RACE

Speed thrillers all day Friday. Night-and-day race starts 10 p. m. Friday. Excitement every minute until 10:30 p. m. Saturday

FIELD SEATS, 75c.   GRAND STAND, $1.50

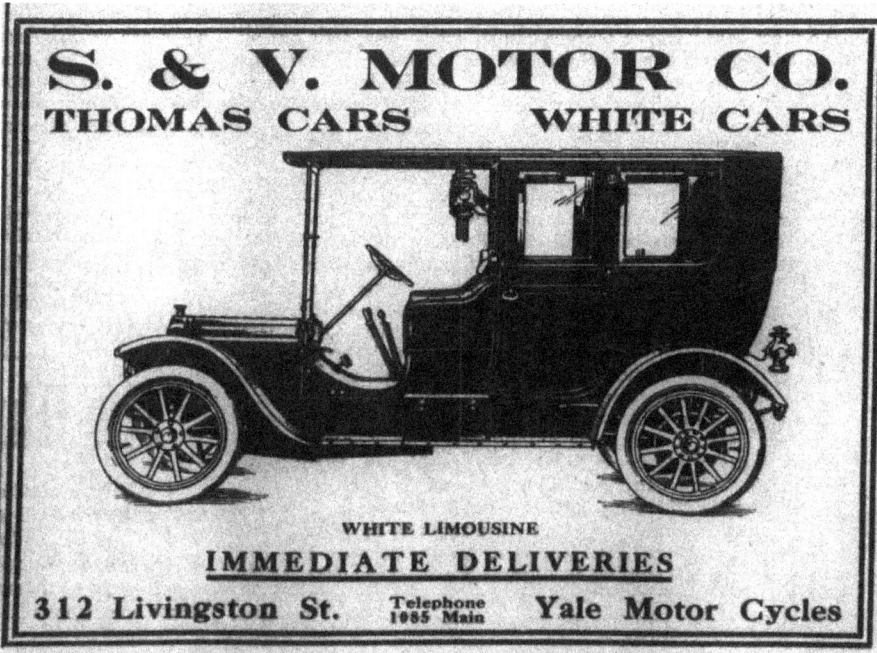

S. & V. MOTOR CO.
THOMAS CARS WHITE CARS

WHITE LIMOUSINE
IMMEDIATE DELIVERIES

312 Livingston St.   Telephone 1985 Main   Yale Motor Cycles

Seventh Season

"30"
"40"
"50"

HALLADAY

$1100
$1250
$1500
$2650

Motor Cars.

Recognized as a high grade product and exceptionally good value.

GRANT SQUARE AUTO CO.
DISTRIBUTERS
1378-82 Bedford Avenue,   BROOKLYN, N. Y.

Model "LC" 14 H. P. $825.

"The Dollar for Dollar Car"

The "Maxwell"

IS THE CONNECTING LINK
BETWEEN MOTORING
AND
SATISFACTION

UNFAILING RELIABILITY

I. C. KIRKHAM

Exclusive Distributer for Long Island
1060 Bedford Ave., cor. Clifton Pl., Brooklyn
Phone: 4300 Bedford

IF YOU WANT A CAR WITH A REPUTATION
IF YOU WANT A CAR WITH ALL PARTS ACCESSIBLE
IF YOU WANT SIMPLICITY, DURABILITY AND ECONOMY
IN SHORT, IF YOU WANT THE BEST THAT MONEY CAN BUY, THEN BUY A

HAYNES

For $2,000
INCLUDES EVERYTHING BUT THE FUEL

1911
REO

A perfect Town Car: It has appointment, style, durability and ease.

The cost of maintenance is extremely low.

The best service you can get on a $2,000 investment.

JOSEPH D. ROURK,

Tel. 3730 Bedford 1001-3 BEDFORD AVENUE. Open Evenings
BROOKLYN AND L. I. DISTRIBUTER

Automobile Owners of Brooklyn and Long Island
Registrations Under the New Callan Law (Continued)

17697 1910 Ford—Henry Blume, 417 Dean St.
17710 1910 Bartholomew—Archibald Nesbit, Bayside, L.I.
17717 1910 Autocar—Jacob Post, Freeport, L.I.
17724 1909 Ford—Chas. E. Hyde, Port Washington, L.I.
17734 1910 Streator—Wm. C. Floeting, 315 Decatur St.
17744 1909 Buick—Chas. Barrucker, 16 Suydam St.
17751 1907 Columbia—Edward D. Morgan, Westbury, L.I.
17753 1910 Buick—Chas. P. Peterson, 1529 East Fourteenth St.
17761 1910 Mercer—F. G. DeWitt, Elmhurst, L.I.
17764 1906 Maxwell—E. C. Granbery, 109 State St.
17767 1909 Jackson—R. E. Williamson, 112 South Portland Ave.
17776 1904 Clement—J. E. Smith, Huntington, L.I.
17778 1910 Ford—Geo. R. Kuhn, M.D., 122 Clinton Ave.
17779 1907 Thomas—H. F. H. Dressel, 209 Howard Avenue.
17786 1909 Cadillac—Edward C. Hopson, 574 Flatbush Ave.
17790 1907 Stevens-Duryea—Raymond Clark, 316 Clinton Ave.
17793 1910 Brush—A. Clark, Flushing, L.I.
17799 1907 Franklin—William Sierks, Rockaway Beach, L.I.
17821 1909 Chalmers-Detroit—E. A. Fleming, M.D., Richmond Hill, L.I.
17822 1907 Ford—F. S. Cantrell, Huntington, L.I.
17828 1909 Chalmers-Detroit—Seaman Brothers, Roslyn, L.I.
17829 1907 Aerocar—J. N. Stewart, Ozone Park, L.I.
17831 1905 National—E. W. Monche, Nineteenth Ave. & Eighty-first St.
17839 1909 Premier—P. H. Bumster, Long Island City, L.I.
17841 1907 Delanney—A. W. Johnson, Oyster Bay, L.I.
17847 1910 Reo—A. P. Matheson, 27 Seventh Ave.
17852 1908 Maxwell—P. J. Van Note, 207 Bay Thirty-fifth St.
17875 1910 Packard—Chas. Lutz & Bro., 185 Harrison Ave.
17881 1909 Hewitt—Kirkman & Son, 215 Water St.
17882 1908 Maxwell—George Keowen, 310 Halsey St.
17889 1907 Haynes—Frank Stamm, 242 Montrose Ave.
17890 1905 Autocar—Henry Worthington, Lawrence, L.I.
17892 1908 Simplex—E. M. Rapalje, 148 Arlington Ave.
17896 1904 Martini—Tyler Morse, Westbury, L.I.
17900 1907 Daimler—Tyler Morse, Westbury, L.I.
17901 1909 Elkhart—Wm. I. Lawnen, Glen Cove, L.I.
17902 1910 Benz—Edward C. Blum, 420 Fulton St.
17906 1909 Regal—V. P. Kerrigan, Hempstead, L.I.
17911 1908 Knox—F. E. Rutand, Huntington, L.I.
17912 1908 Ford—A. J. Tefft, Port Jefferson, L.I.
17923 1909 Pullman—A. B. Gale, 140 Henry St.
17935 1909 Ford—F. J. Weigand, M.D., Richmond Hill, L.I.
17936 1909 Holsman—Wm. H. Trotter, 603 Johnson Ave.
17949 1909 Chalmers-Detroit—A. T. Vance, Port Washington, L.I.
17962 1910 Ford—H. B. Arthur, Smithtown, L.I.
17963 1908 Pierce—Thomas Hoagland, Rockaway, L.I.
17980 1906 Jackson—C. P. Vreeland, E. Ninety-second St. and Ave. M.
17985 1910 Locomobile—Grace Floyd, Greenport, L.I.
17994 1909 Stevens-Duryea—Chas. T. De Bevoise, Hollis, L.I.
17995 1910 Ford—D. S. Whitney, Woodbury, L.I.
18006 1909 Ford—F. B. Vreeland, E. Ninety-second St. & Ave. M.
18007 1910 Buick—Mrs. Harriet Coles, Glen Cove, L.I.
18035 1909 Ford—David Meyer, Sag Harbor, L.I.
18038 1906 Stevens-Duryea—Bedford Garage, Inc., 1293 Bedford Ave.
18042 1910 Hudson—A. L. Smith, Oyster Bay, L.I.
18046 1908 Buick—A. E. Harding, Northport, L.I.
18047 1906 Panhard—Philip Ranche, 137 Ridgewood Ave.
18054 1908 Stoddard—M. J. Scudder, Huntington, L.I.
18056 1908 Stevens-Duryea—D. E. Fondenberger, 192 Nassau Ave.
18057 1907 Packard—Wm. Jos. Duane, 1214 Fifty-fourth St.
18069 1906 Thomas—John Thomas, Woodside, L.I.
18078 1904 Renault—Tyler Morse, Westbury, L.I.
18079 1910 Maxwell—Charles Montagie, Center Moriches, L.I.
18087 1910 Abbott—F. F. Davis, Fort Salonga, L.I.
18109 1907 Cadillac—Van Wyck Hewlett, Woodmere, L.I.
18112 1906 White—Louis Herriman, 98 Stuyvesant Ave.
18122 1910 Chase—New York Telephone Co., 547 Clinton Ave.
18127 1907 Mitchell—Wm. Stoothoff, Baldwin, L.I.
18137 1907 Mitchell—C. H. Walker, Southampton, L.I.
18139 1909 Cadillac—J. K. Worthington, Roslyn, L.I.
18142 1910 Overland—J. D. Cockcroft, Southport, L.I.
18148 1908 Healey—William Dick, 156 South Ninth St.
18154 1909 Pierce-Arrow—Edward Roesler, Great Neck, L.I.
18157 1909 Dayton—L. J. Praeger, Westhampton Beach, L.I.
18161 1909 Jackson—H. I. Van Sicklen, Northport, L.I.
18170 1910 Maxwell—Geo. A. Brush, Huntington, L.I.
18201 1910 Regal—C. C. M. Hoeg, 53 Rutland Road.
18203 1910 Cadillac—L. G. Brimmer, Elmhurst, L.I.
18212 1909 Hupmobile—John Willis, 810 Hemlock St.
18215 1909 Chalmers-Detroit—F. B. Hoffman, Southampton, L.I.
18221 1910 Jackson—Wm. A. Childs, Quogue, L.I.
18224 1906 Clement—J. J. Lowry, Richmond Hill, L.I.
18245 1906 Peerless—W. C. Maclin, 1189 Bergen St.
18248 1909 Regal—C. L. Varrone, Corona, L.I.
18250 1910 Hudson—J. F. Denton, 662 Madison St.
18259 1909 Cadillac—Wm. B. Spencer, Woodhaven, L.I.
18263 1909 Ford—Wm. F. Maass, 582 Glenmore Ave.
18282 1907 Oldsmobile—A. L. Antrum, Arverne, L.I.
18284 1910 Pierce—T. H. Lewis, Great Neck, L.I.
18285 1910 York—J. B. Pitman, Manhasset, L.I.
18289 1907 Thomas—Wm. S. Hofstra, Hempstead, L.I.
18290 1909 Buick—Nassau Lumber Co., Hempstead, L.I.
18299 1910 Saurer—Welz & Zerweck, 1562 Myrtle Ave.
18301 1910 Saurer—Welz & Zerweck, 1562 Myrtle Ave.
18304 1910 Buick—Robert Harrison, E. Patchogue, L.I.
18315 1907 Franklin—James D. Larkin, Southampton, L.I.
18322 1910 Chase—Lowe Brothers, Far Rockaway, L.I.
18324 1910 Baker—Edward Griffin, Oyster Bay, L.I.
18328 1909 Velie—George Russennetter, 1121 Jefferson Ave.
18334 1908 Saurer—Welz & Zerweck, 1562 Myrtle Ave.
18339 1910 Oldsmobile—W. W. De Bevoise, 163 Carlton Ave.

18344 1910 Saurer—Welz & Zerweck, 1562 Myrtle Ave.
18346 1910 Maxwell—J. F. Dreyer, 2021 Dorchester Road.
18347 1909 Hudson—P. D. Rapelje, 948 Belmont Ave.
18352 1905 Oldsmobile—Frank Holbert, 61 Cambridge Place.
18353 1909 Buick—Wm. C. Gray, Patchogue, L.I.
18371 1910 Maxwell—J. A. Duryea, Huntington, L.I.
18373 1909 Buick—J. R. Savage, Jamaica, L.I.
18375 1908 Buick—Welz & Zerweck, 1562 Myrtle Ave.
18378 1910 Velie—Frank Doman, Sea Cliff, L.I.
18387 1910 E.M.F.—Wm. G. Eckstein, Far Rockaway, L.I.
18389 1910 Maxwell—James Pullman, 155 Reid Ave.
18390 1910 Pope—Leo K. Bennett, 382 Fifth Ave.
18393 1908 Atlas—H. C. Bainbridge, 2 Cumberland St.
18395 1910 Maxwell—Mabel Berkheck, Southampton, L.I.
18396 1907 Buick—W. P. Fox, 368 Ovington Ave.
18404 1910 Regal—C. L. Darling, Port Jefferson, L.I.
18411 1909 Jeffery—A. B. Starr, 75 Hawthorne St.
18412 1907 Pope—Jos. Solomons, Flushing, L.I.
18413 1910 Buick—D. W. Moore, 1376 Union St.
18430 1905 Maxwell—Wm. H. Jessup, Jamaica, L.I.
18424 1909 E.M.F.—Wm. H. Shadick, Nassau, L.I.
18442 1907 Craig—Gertrude Wenige, Woodhaven, L.I.
18443 1909 Maxwell—Floyd Grant, Rockville Center, L.I.
18461 1909 Overland—S. D. Sproug, 34 Jefferson Ave.
18462 1908 Maxwell—Marie Weble, Hollis, L.I.
18463 1909 Renault—F. L. Livingston, Bayshore, L.I.
18481 1910 Regal—M. I. Downing, 92 Gates Ave.
18482 1910 Peerless—J. B. Faber, Jamaica, L.I.
18484 1910 Ford—J. J. Strebel, 1048 Decatur St.
18486 1907 Maxwell—M. F. Griffiths, Amityville, L.I.
18501 1910 Hudson—W. J. Hewlett, Cold Spring Harbor, L.I.
18506 1907 Oldsmobile—John Laubden, 48 Clifton Place.
18509 1909 Ford—Wm. D. Outerbridge, Southampton, L.I.
18515 1907 National—L. R. Raeder, 594 Sixth St.
18520 1906 Kobush—James G. Oxnard, St. James, L.I.
18522 1907 Packard—Fred Ihlenburg, 209 Monitor St.
18528 1909 Atlas—J. F. L. Drake, Huntington, L.I.
18531 1910 Maxwell—Wm. C. Benner, Lynbrook, L.I.
18541 1908 Rainier—E. M. Greenfield, 107 Lorimer St.
18556 1909 Chalmers-Detroit—Chas. S. Rienhart, Flushing, L.I.
18560 1909 Buick—Chas. A. Baker, Richmond Hill, L.I.
18570 1907 Rapid—H. E. Ackley, Hempstead, L.I.
18571 1910 National—H. D. Lott, Flushing, L.I.
18574 1908 Ford—Alfred Cohen, Jamaica, L.I.
18579 1908 Autocar—J. H. Plath, 1207 Hancock St.
18580 1909 Marmon—Edward M. Underhill, Glen Cove, L.I.
18585 1909 Packard—Desmond Dunne, 25 Prospect Park West.
18586 1909 Mercer—Chester Weldon, 1633 Fifty-ninth St.
18589 1910 Thomas—Geo. C. Dicrel, Woodhaven, L.I.
18591 1909 Pope—C. A. Pratt, 51 Clark St.
18593 1910 Peerless—F. L. Babcock, Oyster Bay, L.I.
18595 1908 Buick—Korler Bros., Glen Cove, L.I.
18597 1907 Packard—H. I. Pratt, 232 Clinton Ave.
18606 1905 White—Mrs. H. W. Quinn, Richmond Hill, L.I.
18610 1910 Chalmers-Detroit—R. Markey, Jr., Edgemere, L.I.
18611 1908 Haynes—R. E. Meyers, 680 Hancock St.
18628 1906 Peerless—H. I. Pratt, 232 Clinton Ave.
18630 1908 Autocar—E. W. Voorhies, 2188 Ocean Ave.
18634 1907 Stevens-Duryea—J. R. Kevin, 252 Gates Ave.
18639 Autocar—F. L. Babbott, Oyster Bay, L.I.
18645 1909 Buick—A. W. Jagger, M.D., Flushing, L.I.
18700 1910 Moon—J. G. Gasteiger, 91 Bainbridge St.
18703 1910 Hudson—L. W. Fay, 460 Fifteenth St.
18704 1910 Owen—L. L. Cohen, M.D., 760 Bushwick Ave.
18707 1910 Winton—D. W. Hutchinson, 824 Ocean Ave.
18709 1910 Hupmobile—Thos. B. Lynch, Jamaica, L.I.
18711 1909 Mitchell—J. S. Nugent, 97 Hancock St.
18713 1910 Marion—H. C. Murphy, 109 Prospect Place.
18715 1906 Ford—S. Nielsen, 1370 Fifty-sixth St.
18720 1909 Packard—Geo. S. Nicholas, Babylon, L.I.
18727 1910 Cadillac—H. L. Crandall, Freeport, L.I.
18742 1909 Motor—Le Grand L. Benedict, Hempstead, L.I.
18755 1910 Cadillac—Eugene B. Coler, 170 New York Ave.
18761 1905 Panhard—L. T. Haney, 1 Crooke Ave.
18776 1908 Baker—W. S. Grant, 340 Stuyvesant Ave.
18782 1910 White—Thos. E. Smith, Woodhaven, L.I.
18787 1907 Peerless—H. S. Mott, Northport, L.I.
18789 1910 Peerless—Chas. W. Senff, Whitestone, L.I.
18795 1905 Maxwell—S. A. Weber, Hempstead, L.I.
18797 1908 Maxwell—Wm. Rider, Lynbrook, L.I.
18807 1909 Franklin—Samuel B. Althouse, Far Rockaway, L.I.
18819 1909 Stevens-Duryea—Wm. F. Campbell, 394 Clinton Ave.
18821 1910 Buick—Francis Gerber, Sayville, L.I.
18822 1910 Cadillac—Mary F. Harding, 1568 Fulton St.
18823 1907 York—M. A. Furbill, Mattituck, L.I.
18824 1908 Haynes—W. R. Crom, 25 Russell Place.
18828 1905 Thomas—George R. Howell, Southampton, L.I.
18833 1910 Chalmers-Detroit—H. D. Rossen, 10 Maple Court.
18834 1906 Buick—C. A. Hildreth, Southampton, L.I.
18842 1907 Franklin—L. I. Ward, Huntington, L.I.
18849 1909 Stearns—H. C. Bohack, 1090 Greene Ave.
18859 1907 Matheson—G. H. Dorris, North Hempstead, L.I.
18853 1907 Winton—J. J. Bartlett, Greenport, L.I.
18854 1904 Franklin—Y. F. Cornell, Oyster Bay, L.I.
18858 1906 Ford—J. E. Jennings, Water Mill, L.I.
18863 1905 Cadillac—W. B. Bishop, Southampton, L.I.
18865 1909 Crawford—O. M. Jackson, 161 Marlborough Road.
18881 1905 Locomobile—J. L. Lawrence, Lawrence, L.I.
18885 1907 Oldsmobile—Fred W. Brecht, 1580 Flatbush Ave.
18897 1910 Maxwell—W. E. Hulse, Amityville, L.I.
18909 1907 Lozier—R. F. Carman, Garden City, L.I.

(To be continued.)

$1250

A REAL DELIVERY CAR

BEYSTER DETROIT BUSINESS WAGON

This does not mean a pleasure car with a delivery body, but a

COMMERCIAL CAR

from the ground up, 4 cylinder, 4 cycle motor, with Thermo syphon water cooling system; magneto with reserve set of dry cells and storage battery for electric lights; planetary transmission double chain drive; 36-inch wheels, with solid tires; guaranteed for 10,000 miles; speed, 15-20 miles per hour with load.

THEY DELIVER THE GOODS

Give us a chance to demonstrate. **Make us** show you. Call, write or phone for full particulars. Price, with enclosed body, with electric lights, full equipment of tools, **$1,250**, F. O. B., Detroit.

This car is guaranteed for one year. It will do the work of three horses.

THE FARRELL AUTO CO.

159 to 163 REID AVE.

Phone 1147-J Bushwick

Brooklyn and L. I. Distributers

GARFORD "40"—(Formerly Studebaker-Garford)

The first of the beautiful New Model 1911 Garford Enclosed Cars has just been delivered to one of our Brooklyn customers. We will have two more for early delivery. If you are interested in a new Limousine or Landaulet, we would appreciate the opportunity of showing you the new designs.

We have a Studebaker "30" Limousine, taken in trade for a new car, refinished and in fine condition, which we are holding at $2,250.

Carpenter Motor Vehicle Co.
1239-43 Fulton Street, Tel. 1000 Bedford

Photograph by Lazarnick.
MR. ROBERT GUGGENHEIM IN HIS 35-45 RENAULT RACING RUNABOUT.

The "Maxwell"
The World's Champion Endurance Car

$1,100

Fully equipped; 4-cylinder, 25 H.-P.; 104-inch Wheel Base; 32-inch Wheels

Orders Are Now Being Taken for Early Spring Delivery

I. C. KIRKHAM

Tel. 4300 Bedford   1060 BEDFORD AVE.

50 H. P. 114 in. Wheel Base, $2,800
TYPE C

A MOTOR CAR FOR THE DISCRIMINATING.
CORRESPONDENCE AND INSPECTION INVITED.
LEARN WHEREIN "THE DIFFERENCE."
DEMONSTRATIONS BY APPOINTMENT.

GRANT SQUARE AUTOMOBILE COMPANY.

Distributors for Brooklyn and Long Island.   1378 Bedford Ave.

An ideal doctor's car. The first of the 1911 Buick models to appear in Brooklyn.

IMMEDIATE DELIVERY ON A LIMITED NUMBER OF EACH MODEL

22 MODELS FOR 1911

$775 TO $1675

Price $775. Torpedo Body $850
Both Coupe and Runabout Bodies $1050
96 In. Wheelbase, 4 Cylinders, 20-25 H. P.
32 In. Wheels, Magneto, 5 Lamps, Etc.

Price $1250. Touring Car with or without foredoors, same price
112 In. Wheelbase, 4 Cylinders, 30 H. P.
34 In. Wheels, Magneto, 5 Lamps, Etc.

OVER 15,000 CARS IN 1910

Specifications same as Model 46
Price $1000; with Two Bodies $1050

OVER 30,000 CARS FOR 1911

1911 WILL BE Overland YEAR

Price $1600. Touring Car with or without foredoors, same price
118 In. Wheelbase, 7 In. Double Drop Frame
Bosch Magneto, 8 Spark Plugs

Price $1675. Equipped with $75 Warner Speedometer, 2 independent ignition systems, massive steering column, Timken Bearings front and rear, etc.

When you go to buy a Car, "as you will go," you can't afford to forget "*The Overland.*" It is not an assembled car, but is factory made throughout.

Made and guaranteed by the biggest automobile concern in the world, who employ over 5,000 men and have a daily output of over 150 cars. Who can "and will" sell automobiles in 1911 for less than any other concern could even build them, "if they built them as good as Overlands." The Cost Cut 28% Overland prices are due to the use of modern automatic machinery—acres upon acres of it. Also to enormous production. Over $3,000,000 has, to-date, been invested in the best of labor saving equipment. We have thus cut the cost of Overland 28 per cent. in the past two years—an average of $300 per car. At the same time we have secured such exactness as could never be secured in the old ways. And we have made every similar part interchangeable.
It is thus that we undersell (by far) every other maker who puts out a high-grade car.

Write for Catalogues   Prices F. O. B. Factory

EASTERN DISTRIBUTORS:

OVERLAND SALES CO. OF N. Y.

New York Office:
1599 Broadway
Tel. { 5741 / 5742 / 5743 } Bryant

C. T. SILVER

62 Flatbush Ave., B'klyn
Tel. 379 Main

AGENTS GET BUSY

N. Y. Salesroom open evenings until 12

The MAJA Car

Pronounced "My-yah"

THE SISTER OF MERCEDES

The Daimler Motoren Gesellschaft, builders of the Mercedes car, have turned over their entire Austrian works to the manufacture of a new and improved car, called MAJA (pronounced "My-yah"), embodying all their best features, greatly simplified, refined, improved, where improvement was possible, and have made Maja their leading car in order to market it themselves, avoiding all the commissions of middlemen and agents, and the mistakes in marketing their product in the past.

The output of the Daimler Works has been generally recognized as representing the leading and foremost automobile productions of the world. There have been vastly greater sums paid for Daimler productions than for any other car in the world, and now, through the medium of the Maja car, these great creations are within reach of every automobile buyer.

Maja motors are much more powerful than anything heretofore produced of equivalent size. They are vastly simpler than anything Europe has ever offered, and no motor in the world can approach Maja for absolute silence.

Maja costs more to manufacture than any previous model; but, by the economies of direct marketing, is placed in easy reach of every user of good automobiles.

The Maja car will pass through the hands of no middlemen or agents, and with the exception of Germany and Austria will be sold direct by the Maja Company, Ltd., with offices and branches in London, Paris, Stuttgart, Hamburg, St. Petersburg and New York.

Full particulars from the American branch of the Maja Company, Ltd., 230 West 58th St., New York.
Telephone 1393 Columbus.

Herr Jellinek named his first famous success in honor of one of his daughters, "Mercedes." His newest creation, embodying every improvement that time and trial have wrought under his skill—his latest and crowning achievement—he has named for another daughter, Fraulein Maja Jellinek, the "Maja" car, (pronounced "My-yah").

The Maja car is therefore in fact and in name the full sister of Mercedes

You are invited to attend "Maja's" debut at the Importers' Automobile Salon from December 28th to January 7th at Madison Square Garden.

As is befitting, "Maja" has been assigned the leading position in the Show, immediately at the left of the main entrance. Space A-1.

A RENAULT 14-20 H. P. CAR, BODY BY KELLNER.

ONE OF THE 1908 ISOTTA FRASCHINI CARS.

A 24 H. P. CLEMENT-BAYARD LANDAULET.

THE FIRST 28-35 H. P. MAJA—"THE SISTER OF MERCEDES."

Some New Models of Foreign Cars at the Importers' Salon in Madison Square Garden.

THE INCOMPARABLE
WHITE
THE CAR FOR SERVICE

EXCLUSIVE FEATURES OF THE WHITE LIMOUSINE

The exclusive White quality of absolute noiselessness of operation is of particular advantage in a limousine because, in a car with a closed body, any noise made by the mechanism is even more noticeable and annoying than in an open vehicle.

Another exclusive White quality, namely, genuine flexibility of control, permits of the machine being guided safely and speedily through the crowded city streets. The speed of the White may be accommodated to the exigencies of street traffic without any changing of gears, jerky starts or the embarrassing and sometimes dangerous "stalling" of the engine.

As regards graceful lines and luxuriousness of equipment and finish, the White limousine must be seen to be appreciated.

Let us show you the unequalled luxury and comfort of the White Limousine.

THE WHITE COMPANY
Broadway at 62nd Street, New York City

"MODEL T" $1000.00

CADILLAC A REAL Christmas Present

Jos. D. Rourk

Sole Agent Brooklyn and Long Island

Telephone 3730 Bedford.   **1001=3 Bedford Ave.**

MERCANTILE MOTOR CAR CO. (Inc.)

Successors to Long Island Motor Car Co.   Telephone. 2538 Prospect

General Machinists

FIRST-CLASS CARS
Fully Equipped for Renting

Sole Agents for Brooklyn
and Long Island of
"ALUMUNITE"
The Only Aluminum Solder

Automobiles Stored

And Repaired

Dealers in Second-hand
Automobiles

368 Cumberland Street,   Brooklyn, N. Y.

AUTOMOBILE SCHOOL FOR OWNERS, PROSPECTIVE OWNERS AND CHAUFFEURS

Lectures, shop work and road lessons. New class now forming, limited to ten.

BEDFORD BRANCH, Y.M.C.A.
1121-25 BEDFORD AVENUE,
Corner Monroe St., Brooklyn.

Lancia in his 110 H. P. FIAT in the Vanderbilt Cup Race
"The Fastest Car in the World."—*New York Herald*

We are receiving each week two or three of the famous FIATS and are therefore prepared to make prompt delivery of these cars, completely equipped and ready for the road. "FIAT" cars may be had with either closed or open bodies, the work of the very best European and American body designers.

HOLLANDER & TANGEMAN
3 and 5 West 45th Street   -   -   -   New York

Sole American Agents.   Licensed Importers Under Selden Patent.
Fiat Cars will be exhibited only at Madison Square Garden Automobile Show

Livery greatcoats that are all wool, and so keep the coachman warm as toast.

As the coachman ought to be.

Descriptive folder on application

ROGERS, PEET & CO.
258—842—1260 Broadway
(3 Stores)
NEW YORK

Rambler Cars of steady service for 1906

IN SELECTING our line for the coming season it was early decided that Surreys, Types One and Two, were beyond any question of retirement and that but few improvements were possible. Therefore, these models with some slight alterations will be continued. To these are added Type Three, which is practically an elaboration of Type One, the power plant and chassis remaining the same, except somewhat lengthened to accommodate a larger and longer body.

These models are equipped with the tried and proven Rambler power plant, comprising our double opposed motor and planetary pattern transmitting gear, and are too well known to the trade and public to require extended description at this time.

Prices: Type One $1,200, Type Two $1,650, Type Three $1,350, all with full equipment of lamps, horns, tools, etc.

"*The Latest of the Ramblers,*" the strictly 1906 product, comprises four models. Model Fourteen is a modern medium weight touring car equipped with a four-cylinder vertical motor 20-25 horse power, with sliding type transmitting gear, giving three forward speeds and reverse.

Final drive is by propeller shaft and bevel gear to the differential on the rear axle.

A notable feature is the method of connecting and bracing this shaft in which the universal joint is at the forward end and is entirely enclosed, running in an oil bath.

The external design is along most modern lines with a wheel base of 106 inches.

The selling price of this model is $1,750 with complete equipment.

Model Fifteen is a heavier car with similar but more powerful equipment, the motor being 35-40 horse power and the final drive by individual chain to each rear wheel. The body is practically the same as in Type Fourteen but longer and larger, the wheel base being 112 inches.

Model Sixteen is a most luxuriously appointed Limousine on the Model Fifteen chassis, selling at $3,500.

The Rambler runabout for 1906 is a fitting heir to the reputation gained by the earlier Ramblers of this type. It is equipped with a double opposed motor of 10-12 horse power, placed longitudinally in the frame and driving through the Rambler planetary gear. As a Runabout with 3-inch tires it will sell at $800 and with detachable tonneau and 3 1-2 inch tires at $950.

Catalogue and full descriptive matter will be mailed upon request, but a careful personal examination of these cars at our various branches and agencies, will convince you that whatever may be your requirement, in service or price, the Rambler is the car you need.

Thos. B. Jeffery & Co.

Main Office and Factory, : : Kenosha, Wis., U. S. A.

Branches:

Boston, 145 Columbus Avenue   Chicago, 302-304 Wabash Avenue   Philadelphia, 242 N. Broad Street
Milwaukee, 457-459 Broadway   San Francisco, 10th and Market Streets
New York Agency, 134 W. 38th Street   Representatives in all leading cities

We are Ready

TO MAKE AN APPOINTMENT
WITH YOU

To Demonstrate

THE QUIET-MILE-A-MINUTE
1906--35-40 H. P.

Pope-Toledo

The highest type of advanced automobile construction. Strong, powerful, simple and exceedingly easy to control. Speed 4 miles to 60 miles on direct drive high speed

A. G. SOUTHWORTH

342 Flatbush Avenue   **10 Clinton Street**

1911 *Buick* CLOSED CARS

This is the season for powerful closed cars.

Purchase to-day a 1911 Buick limousine which is equal in style, size, material, workmanship, design, finish, detail, comfort, A. L. A. M. horse power, economy of upkeep, and service to the popular $5,400 and $5,600 closed cars. You will save money enough to pay the operating expenses for a year as follows:

Chauffeur, steady employment year round	$1,040.00
N. Y. City Garage bill @ $40.00 month year round	480.00
Maximum tire bills, including four extra 36x4½ inch rear and two 36x4 inch front shoes and tubes	400.00
Gasoline for 7,500 miles @ $.20 gallon, allowing gallon for only ten miles	150.00
Cylinder oil for 7,500 miles @ $.70 gallon, allowing one gallon for every hundred miles	52.50
Transmission oil, cup grease, waste charging batteries	25.00
Liberal total based on maximum expenses	$2,147.50

In view of these concrete facts, and our branch house guarantee of service, vouched for by 62,000 satisfied customers, can you—a live up-to-date thinking business man, in these days when guaranteed service and economy counts—afford to purchase elsewhere? Investigate.

A closed car is a most gratifying and practical Christmas present—a necessity for comfort and modern social requirements.

BUICK MOTOR COMPANY

NEW YORK
Broadway at 55th St.

BROOKLYN
42 Flatbush Ave.

NEWARK
222 Halsey St.

When real Winter comes a really warm greatcoat is a necessity for even the hardiest of coachmen.

Our greatcoats are all lined to the bottom of the skirt with an all wool lining weighing 30 ounces to the yard.

Made of American or imported English boxcloth $38 to $100.

ROGERS, PEET & CO.
258—842—1260 Broadway
(3 Stores)
NEW YORK

A GROUP OF DETROIT WOMEN IN A NORTHERN MOTOR CAR.

From the poster by E. Montaut

Szisz Sur Voiture Renault Frères, Gagnant du Grand Prix de l'Automobile Club de France.

QUINBY

ALUMINUM BODIES

LIMOUSINE,
LANDAULET,
TOURING,
RUNABOUT.

COMPLETELY EQUIPPED
MOTOR REPAIR
DEPARTMENT.

PANHARD AND SIMPLEX CARS

Early Delivery

J. M. QUINBY & CO.

NEWARK, N. J.
Adjacent Lackawanna Station.

RENAULT
"The Car"

Style   Silence   Speed   Simplicity

All Over the World the Renault is today the Acknowledged Standard of Automobile perfection

In the famous 24-hour race at Morris Park the Renault, with a stock car and without making a single adjustment, broke all world's 24-hour records for a single car in competition on a mile track, making 1,079 miles, defeating the nearest competitor by 107 miles

RENAULT PRICES:

50-60 H.P. 6 CYLINDER CHASSIS	$8,250	20-30 H.P. 4 CYLINDER TOURING	6,250
35-45 H.P. 4 CYLINDER RUNABOUT	7,250	20-30 H.P. 4 CYLINDER LIMOUSINE OR LANDAULET	6,750
35-45 H.P. 4 CYLINDER TOURING	7,750	14-20 H.P. 4 CYLINDER CHASSIS	4,250
35-45 H.P. 4 CYLINDER LIMOUSINE OR LANDAULET	8,250	10-14 H.P. 4 CYLINDER TAXIMETER CAB	3,950
20-30 H.P. 4 CYLINDER RUNABOUT	5,750	8-10 H.P. 2 CYLINDER TAXIMETER CAB	2,750

DO YOU BELIEVE IN RELIABILITY?

These gentlemen do. They have ordered 35-45 H. P. Runabouts, duplicates of record-breaker:

W. K. Vanderbilt, Jr.	L. S. Thompson	R. J. Collier	H. P. Whitney
E. R. Thomas	E. F. Hutton	Roy Rainey	Payne Whitney
Robert Guggenheim	Willis MacCormick	F. W. Savin	W. E. Dodge
	J. L. Livermore		G. A. Ellis

These men know about cars—why not follow them?

RENAULT FRÈRES SELLING BRANCH

PAUL LACROIX, General Manager   Telephone, 3004 Columbus

57th Street and Broadway, New York City

RENAULTS will be shown only at Importers' Salon, Madison Square Garden, Dec. 28th to Jan. 4th

Why the Ful-Floteing Seat Makes This Motorcycle Comfortable

THE ordinary motorcycle equipped with a rigid unadjustable leaf or coil spring that must be strong enough to support a 300 lb. man or break, makes mighty uncomfortable riding for a rider of lesser weight, just the same as a light auto equipped with the heavy, stiff springs used on auto trucks would make riding for its occupants anything but comfortable. The adjustability of the springs in the FUL-FLOTEING SEAT, an exclusive patented feature of the

HARLEY-DAVIDSON

makes the Harley-Davidson the one motorcycle that is absolutely comfortable for every rider, regardless of weight. A comfort device, to be worthy of its name, must not only protect the rider from the jars and jolts of rough roads, but must also overcome an equally objectionable feature, the rebound, so common to the ordinary motorcycle.

The FUL-FLOTEING SEAT of the Harley-Davidson virtually floats or carries the weight of the rider between two concealed compressed springs. One of these springs assimilates the jolts and jars due to bumps; the other absorbs the rebound. Thus the rider feels no jars, vibrations or rebound, but floats along in comfort.

The strength of these springs can be adjusted by a few turns of a tension nut so that they will operate perfectly under the weight of the rider. Thus the rider of a Harley-Davidson experiences pleasure and comfort going over the roughest road as he does when riding on the smoothest boulevard.

When you purchase a motorcycle—get a comfortable one with an adjustable FUL-FLOTEING SEAT—a Harley-Davidson. Ask the nearest Harley-Davidson dealer for a demonstration or write for catalog.

HARLEY-DAVIDSON MOTOR COMPANY
PRODUCERS OF HIGH GRADE MOTORCYCLES FOR OVER ELEVEN YEARS
331 B Street   MILWAUKEE, WISCONSIN

THE INCOMPARABLE
WHITE
THE CAR FOR SERVICE

EXCLUSIVE FEATURES OF THE WHITE LIMOUSINE

The exclusive White quality of absolute noiselessness of operation is of particular advantage in a limousine because, in a car with a closed body, any noise made by the mechanism is even more noticeable and annoying than in an open vehicle.

Another exclusive White quality, namely, genuine flexibility of control, permits of the machine being guided safely and speedily through the crowded city streets. The speed of the White may be accommodated to the exigencies of street traffic without any changing of gears, jerky starts or the embarrassing and sometimes dangerous "stalling" of the engine.

As regards graceful lines and luxuriousness of equipment and finish, the White limousine must be seen to be appreciated.

Let us show you the unequalled luxury and comfort of the White Limousine.

THE WHITE COMPANY
Broadway at 62nd Street, New York City

Locomobile

Eleven years' of motor car manufacture establishes the famous Locomobile in the highest plane of refinement and development. In purchasing a Locomobile you secure a car of unquestioned high reputation, a thoroughly safe selection. You also secure the benefits derived from doing business with a long established and experienced organization.

The "30" Locomobile, Shaft Drive
The "40" Locomobile, Chain Drive
TOURING CARS, ROADSTERS, LIMOUSINES, LANDAULETS

THE I. S. REMSON M'F'G. CO.
BROOKLYN AGENTS
754-760 Bedford Ave.　　　Tel. 2826 WmBurgh,　　　Brooklyn.
W. H. KOUWENHOVEN, MANAGER AUTO DEPARTMENT.

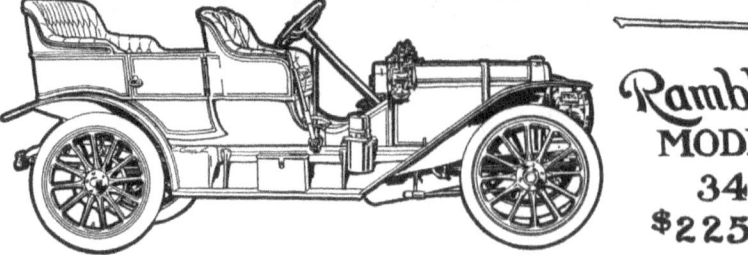

Rambler
MODEL 34
$2250

It is the Practical Things that Count

IN the recent three-day 600-mile Reliability Run of the Chicago Motor Club a Rambler stock car with no special preparation or equipment made every control on time without an instant's delay or attention and finished in perfect condition.

A "hypertechnical" committee deprived us of a perfect score on account of a damaged tail lamp and loosened speedometer bracket, but the fact remains that this car underwent the most severe test ever devised without a seal being broken or the tool bag opened, and was at the finish in the same perfect condition as at the start.

Thus was the Rambler again proven

THE CAR OF STEADY SERVICE

If you want a car that is Right at a price that is Reasonable with the backing of one of the most powerful companies in the industry, see the Rambler line for 1908.

Two touring cars and a roadster at $1,400 and $2,250.

Catalog 34 upon request

THOMAS B. JEFFERY & COMPANY
MAIN OFFICE AND FACTORY, KENOSHA, WIS.

New York Representatives: HOMAN & SCHULZ, 38-40 West 62d Street

Pope Hartford "30" Touring Car Price, $2,750

Direct Factory Representative for all POPE Cars, for Brooklyn and Long Island

For three years the **POPE HARTFORD** car has been steadily gaining in well deserved popularity. You will make a mistake if you buy a car before trying the 1908 HARTFORD "30." Just ring Prospect three-one-one-four. The car will meet you at the door.

JOHN W. SUTTON

A. W. BLANCHARD, Mgr.   342 Flatbush Ave.

THE 1908, 25 H.P. HOL-TAN ROADSTER; THE DESIGNER, LOUIS P. MOOERS, AT THE WHEEL

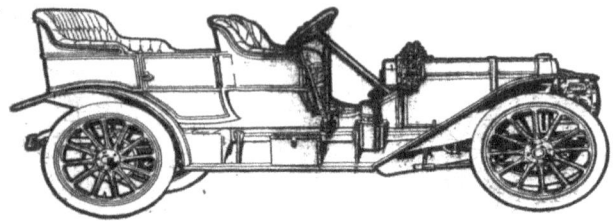

Rambler MODEL 34 $2250

It is the Practical Things that Count

IN the recent three-day 600-mile Reliability Run of the Chicago Motor Club a Rambler stock car with no special preparation or equipment made every control on time without an instant's delay or attention and finished in perfect condition.

A "hypertechnical" committee deprived us of a perfect score on account of a damaged tail lamp and loosened speedometer bracket, but the fact remains that this car underwent the most severe test ever devised without a seal being broken or the tool bag opened, and was at the finish in the same perfect condition as at the start.

Thus was the Rambler again proven

THE CAR OF STEADY SERVICE

If you want a car that is Right at a price that is Reasonable with the backing of one of the most powerful companies in the industry, see the Rambler line for 1908.

Two touring cars and a roadster at $1,400 and $2,250.

Catalog 34 upon request

THOMAS B. JEFFERY & COMPANY
MAIN OFFICE AND FACTORY, KENOSHA, WIS.

New York Representatives: HOMAN & SCHULZ, 38-40 West 62d Street

THE HEALY RAPID REMOVABLE RIM

NEW PLAZA GARAGE

WILLIAM SINN GROVER, Manager
Union Street and 9th Avenue, Brooklyn.

Cars to hire $3 and $4 per hour. Special rates by day or week, Weddings and Theatres.

ALL LARGE HANDSOME CARS

Dead Storage $5, 6, 7.50 p. w. Tel. 969 & 3414 Prospect

AUTOMOBILE SCHOOL FOR OWNERS, PROSPECTIVE OWNERS AND CHAUFFEURS

Lectures, shop work and road lessons. New class now forming, limited to ten.

BEDFORD BRANCH, Y.M.C.A.
1121-25 BEDFORD AVENUE,
Corner Monroe St., Brooklyn.

"She tried to cure his whiskey habit

AUTOMOBILE CLOTHING
FOR OWNER AND CHAUFFEUR

Suits, Overcoats, Hats and Caps, Fur, Fur Trimmed and Fur Lined Coats for men and women

A. J. NUTTING & CO., Inc.
FULTON and SMITH STS.   **BROOKLYN.**

Motor Apparel Shop
Holiday Suggestions for the Motorist

Fur Coats	Flower Holders
Fur Lined Coats	Clocks
Fur Hats	Luncheon Baskets
Fur Caps	Automobile Trunks
Fur Robes	Hampers
Fur Gloves	Fitted Bags
Raincoats	Steamer Rugs
Touring Coats	Silver Flasks
Angora Jackets	Thermos Bottles
Sweaters	First Aid Cases
Bonnets	Drinking Cups
Veils	Vanity Bags
Gloves	Pillows in Leather Cases
Motor Scarfs	Foot Warmers
Mufflers	Dunhill Robes
Polo Coats	Plush Robes
Leather Coats	Air Cushions

And Many Other Novelties

Fall and Winter Catalog sent postpaid on request.

Fox, Stiefel & Co. FIFTH AVE. & 34th St. N.Y.
Opposite the Waldorf-Astoria

The Master Car
OF THE
Master Craft

Completely equipped with everything but the fuel and trouble

JOS. D. ROURK
1001-3 Bedford Avenue
Brooklyn and L. I. Distributor

Tel. Bedford 3730.

CEDRINO IN THE "FIAT" JUNIOR

(Which holds the world's middle-weight record) at Ormond.

The Locomobile

ANNOUNCEMENT, 1911
THE "30" SHAFT-DRIVE—FOUR CYLINDERS
THE "48" SHAFT-DRIVE—SIX CYLINDERS

High Tension Dual Ignition System on both models. Four speed selective transmission. A wide range of the latest body styles—either with or without front doors—can be supplied. Touring, Baby Tonneau, Runabout, Torpedo, Limousine and Landaulet. Finished in any color scheme desired by the purchaser.

Complete Information Furnished on Request

THE I. S. REMSON M'F'G CO.
BROOKLYN AGENTS
754-760 Bedford Ave.   Tel. 2826 WmBurgh.   Brooklyn.
W. H. KOUWENHOVEN, MANAGER AUTO DEPARTMENT.

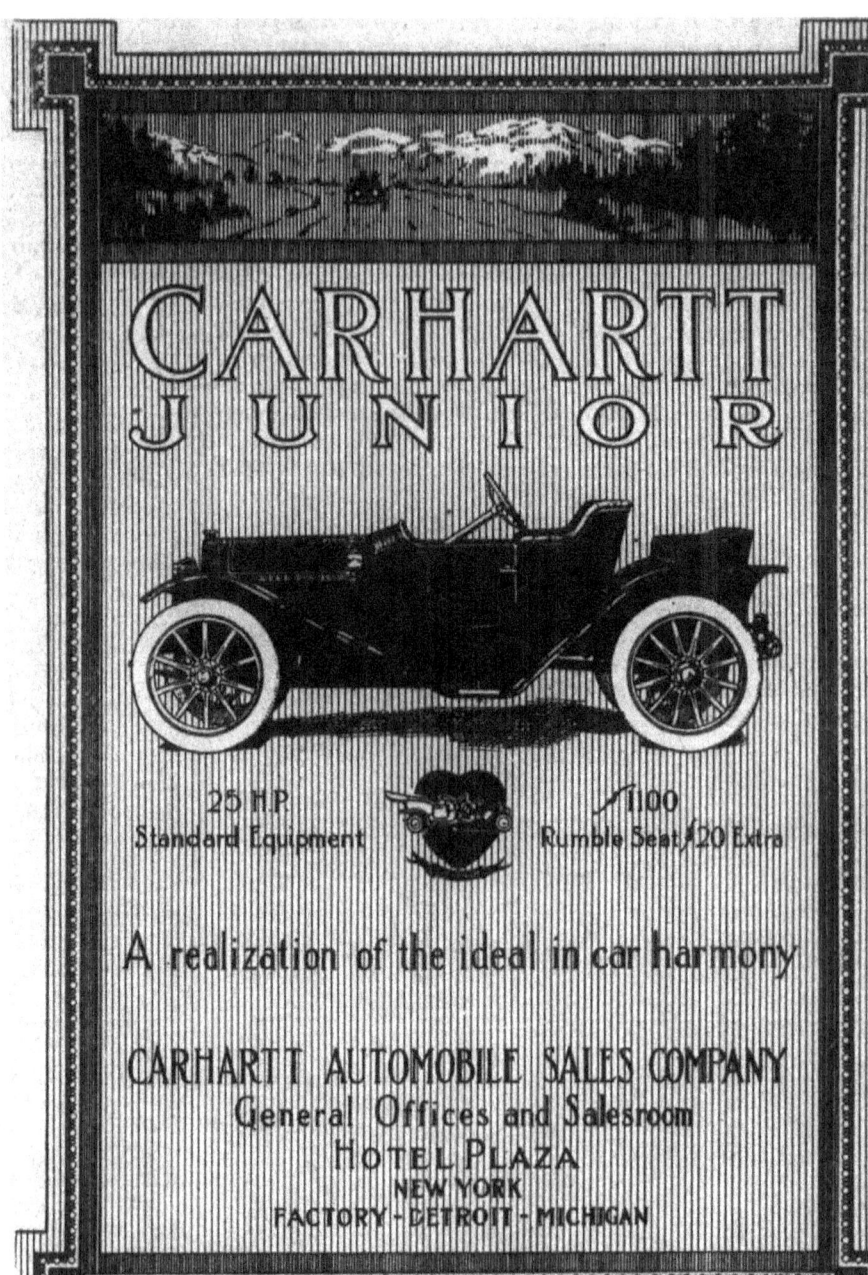

SILENCE
COMFORT

TOWN CAR

20 H.-P. Town Car

THE demand for a lighter car, which is especially adapted to city and suburban use, is admirably met in the new Peerless Town Car. While retaining all the luxurious appointments which have given such distinction to the Peerless cars, this new model is designed for less strenuous service, where modified power, lighter construction, shorter wheel base, and left-hand drive, are points of convenience and desirability.

This new car may now be seen at our Salesrooms. We are prepared to accept a limited number of orders for immediate delivery.

THE PEERLESS MOTOR CAR CO. OF N. Y.
1760 Broadway, at 57th Street
PEERLESS GARAGE & SALES CO.
1525 Bedford Ave., near Eastern Parkway
Licensed under Selden Patent

The  30 H. P.

MODEL 17 TOURING CAR

4 cylinders cast in pairs, 4½ x 5 inches. With magneto and 5 lamps.

PRICE $1,750.

Model 17 Buick

This Model and the Model 16 four passenger Tourabout mounted on the same chassis, are without exception the **strongest, safest and easiest riding** cars sold at a moderate price. They will last as long, cost less to operate, and bring a better percentage of the original price when sold second-hand than more expensive cars. Therefore, they are a better investment.

All Buick cars are simple in construction and easy to operate. We give purchasers thorough instruction. The Buick policy protects every owner.

BUICK MOTOR COMPANY

NEW YORK	BROOKLYN	NEWARK
55th St. and Broadway	42 Flatbush Ave.	222 Halsey Street

Reliability, Luxury, Refinement
Studebaker "40"

STUDEBAKER ELECTRICS
Landaulets, Coupes, Victorias
(Interior Driven)

"For the Family's use"

CARPENTER MOTOR VEHICLE COMPANY
BROOKLYN AGENTS
1239-41-43 FULTON STREET
Telephone 300 Bedford for demonstration.   OPEN EVENINGS

Don't Take the Other Fellow's Dust

"40" $1,800

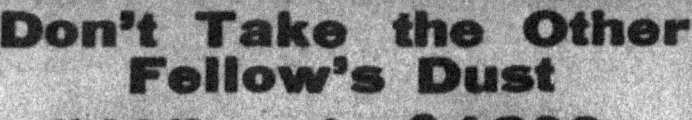

Quality Cars
Touring -- Toy Tonneau -- Roadster

J. MORA BOYLE,   BROADWAY,
Pres.   at 61st St.

HOTELS

I. M. ALLEN COMPANY

INSTANT DELIVERY.

(MOTOR CARS OF 1910)
116 South Portland Av.,
Between Fulton St., and Hanson Place.
TELEPHONE, 4026—PROSPECT.
BROOKLYN, N. Y.
Manufactured By
STEVENS-DURYEA CO., CHICOPEE FALLS, MASS.
MEMBERS A. L. A. M.

Kaiser Haus

1155-1157 Fulton St., at Franklin Ave.
Table d'Hote Lunch 40c., Dinner 60c.,
Holidays and Sundays $1.00
Accommodations for Theater parties, Banquets
Stag, Beefsteak parties and Thumb Bits.
Private alleys for Ladies' Bowling Clubs.
GEO. BOEMERMANN

TRUFFAULT-HARTFORD
SHOCK ABSORBER

Indispensable for Comfortable Motoring

Hartford Suspension Co.
173 Bay St., Jersey City, N.J.
New York—212 W. 88th St.
Boston—219 Columbus Ave.

THE CLEVELAND PATHFINDING CAR ON ROAD LEADING INTO DAYTONA, FLORIDA—AS DESCRIBED ON OPPOSITE PAGE.

DETROIT CHILDREN MOTORING IN A PUNGS-FINCH LIGHT TOURING CAR.

1908
Type XVII Pope-Toledo

JUST ARRIVED

¶ The best built and finest equipped and finished Car in America to-day. Don't buy before seeing it.

JOHN W. SUTTON,
342 FLATBUSH AVE.

10 Clinton St.   811 Union St.

A. W. BLANCHARD, Mgr.

Also

Pope-Hartford
Pope-Tribune
Pope-Waverly Electric

The Incomparable
WHITE

The Car for Service

WHITE RELIABLILITY AND LOW
GASOLINE CONSUMPTION
AGAIN DEMONSTRATED

The silent, swift and sturdy White steam car starts on its sixth year of successful service with a sweeping victory in an important competition. The Los Angeles-San Diego endurance contest, held January 25th-26th, was won by a White steamer carrying five passengers and 150 lbs of baggage. The wining White, driven by Charles A. Hawkins, made a perfect score—1,000 points. One other car, of 40 per cent greater cost than the White, also made a perfect score, but as Mr. Hawkins had used but *seventeen gallons* of gasoline, as compared with *twenty gallons* used by his advesary, Mr. Hawkins's White was declared the winner, in accordance with the rules of the contest. Three other Whites, driven by private owners, participated and received first class certificates, losing but one, two and three points, respectively.

The results of the above competition, except for differences in dates and places, read much like the summaries of every other reliability trial which has been held, starting with the New York-Rochester run of 1901.

WRITE FOR LITERATURE

WHITE Sewing **CO.** Cleveland
Machine Ohio

White Garage, 1380 Bedford Ave.
Brooklyn.

Waiting for you.

With everything any liveried

servant wears.

Indoors or out.

ROGERS PEET & CO.
258—842—1260 Broadway
(3 Stores)
NEW YORK

Priestley's

ENGLISH
TUSSAH ROYAL

The New Worsted and Mohair Dress Fabric

FOR SPRING AND SUMMER WEAR

Beautiful Draping Qualities
Brilliant Silky Appearance
Sheds Dust and Will Not Wrinkle.

For sale by the leading Dry Goods Stores in Black and Colors in all the fashionable weaves.

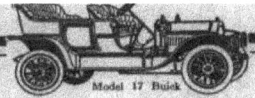

The  30 H.P.

Model 17 Touring Car

4 Cylinders, Cast in Pairs, 4½x5 Inches, with Magneto and 5 Lamps

PRICE $1,750

WE have just received the following letter from Mrs. James N. Dillard about her Buick Car. It is like hundreds of others received daily from satisfied customers.

Buick Motor Co., 42 Flatbush Ave.,
    Brooklyn, N. Y.:
    Gentlemen—I am glad to express my appreciation of the Buick Car. My father, Captain J. H. Wood, purchased a Buick last year, and I, being extremely fond of driving, having driven fine horses from my girlhood, I found myself equally fond of driving the Motor Car. Hence I have driven the Buick almost exclusively since it was purchased and find it to be the readiest, most responsive, easiest managed and reliable car I have ever had anything to do with. Furthermore, the uniform courtesy and kindness of the company is commendable and much appreciated. We like the business policy of the Company and we like the Buick.   Yours respectfully,
    (Signed)   MRS. JAMES N. DILLARD.

Our Model 19 Demonstrator Has Just Arrived—PRICE $1,400

 Policy

IT has always been the policy of the Buick Motor Company to take proper care of its customers. In 1910 this will be the highest aim of our entire organization. With this thought in mind we solicit your valued trade.

We Demonstrate

 CARS

On Miller Avenue Hill   Ask Others To Do Likewise

We Guarantee That the Cars We Deliver Will Climb Miller Avenue Hill on High Gear

BUICK MOTOR COMPANY

| NEW YORK | BROOKLYN | NEWARK |
| Broadway & 55th St. | 42 Flatbush Ave. | 212 Halsey St. |

KNOX

It is attention to details that marks the difference between a high and a low grade car; the brain cost against the material cost. You will find both in the

Knox Waterless

Price, $2,500

25-30 H. P. — 102" wheel base — 32" x 4" wheels — weight 2250 lbs. — automatic oiling — direct cooling — the troubless car of 1907. Let us send you our 1907 catalogue; it illustrates and describes this and four other models.

Knox Automobile Company

Members Association Licensed Automobile Manufacturers
SPRINGFIELD, MASS.

WINTON

Next year almost every high-grade car will have Offset Cylinders and the Multiple Disc Clutch. If you don't care to wait a year you can get them right now in the Winton Model M. Another Winton feature is a Carburetor that, for power and speed, distances the field.

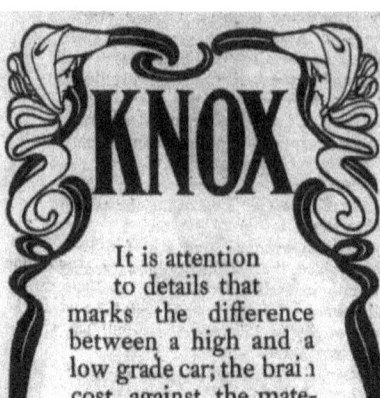

CARLSON AUTOMOBILE CO.
1060 Bedford Avenue, Brooklyn

POPE AUTOMOBILES

TYPE XV
TOLEDO
50 H. P. TOURING CAR OR runabout, $4,250.

The
CHROME STEEL FLYER
"MILE-A-MINUTE"

Fully guaranteed for one year
A. G. SOUTHWORTH & CO.
Inc.
New York   Brooklyn
1733 Broadway   342 Flatbush Ave.

Ice, Mud and Roughly Frozen Roads serve only to emphasize the exclusive excellence of

Diamond
WRAPPED TREAD TIRES

Mr. R. G. Kelsey said

at the conclusion of his recent midwinter run, in a Matheson stock car, from New York to Chicago:

"I will confess that at one time I was a staunch exponent of imported tires; but I am absolutely converted to Diamond Tires as being superior to anything with which I have ever had any of my machines equipped."

The Diamond Rubber Company
AKRON, OHIO

Photograph by Lazarnick.
THE RECENT IMPORTERS' SALON: PART OF THE MAJA EXHIBIT.

Model 17 Buick

The *Buick*
30 H.P.
Model 17 Touring Car

4 Cylinders, Cast in Pairs, 4½x5 Inches, with Magneto and 5 Lamps
PRICE $1,750

WE have just received the following letter from Mrs. James N. Dillard about her Buick Car. It is like hundreds of others received daily from satisfied customers.

Buick Motor Co., 42 Flatbush Ave.,
  Brooklyn, N. Y.:
  Gentlemen—I am glad to express my appreciation of the Buick Car. My father, Captain J. H. Wood, purchased a Buick last year, and I, being extremely fond of driving, having driven fine horses from my girlhood, I found myself equally fond of driving the Motor Car. Hence I have driven the Buick almost exclusively since it was purchased and find it to be the readiest, most responsive, easiest managed and reliable car I have ever had anything to do with. Furthermore, the uniform courtesy and kindness of the company is commendable and much appreciated. We like the business policy of the Company and we like the Buick. Yours respectfully,
(Signed)   MRS. JAMES N. DILLARD.

Our Model 19 Demonstrator Has Just Arrived—PRICE $1,400

Policy

IT has always been the policy of the Buick Motor Company to take proper care of its customers. In 1910 this will be the highest aim of our entire organization. With this thought in mind we solicit your valued trade.

We Demonstrate

CARS

On Miller Avenue Hill   Ask Others To Do Likewise

We Guarantee That the Cars We Deliver Will Climb Miller Avenue Hill on High Gear

BUICK MOTOR COMPANY

NEW YORK	BROOKLYN	NEWARK
Broadway & 55th St.	42 Flatbush Ave.	222 Halsey St.

Model 16 Buick

PENNSYLVANIA

There is nothing better than THE BEST

Built Right
Sold Right

✦ ✦ THERE ARE NO HILLS ✦ ✦

50 H.P. 114 In. W.B.
$2,800

P. S.—The gentleman from Missouri will please call

GRANT SQUARE AUTOMOBILE COMPANY
BROOKLYN & L. I. DISTRIBUTORS
1378 BEDFORD AVENUE, - BROOKLYN, N. Y.

Photographed for Brooklyn Life by F. A. Walter.
The new fireproof salesroom, repair shop and garage of the Mitchell Motor Co., at 24 Kosciusko Street.

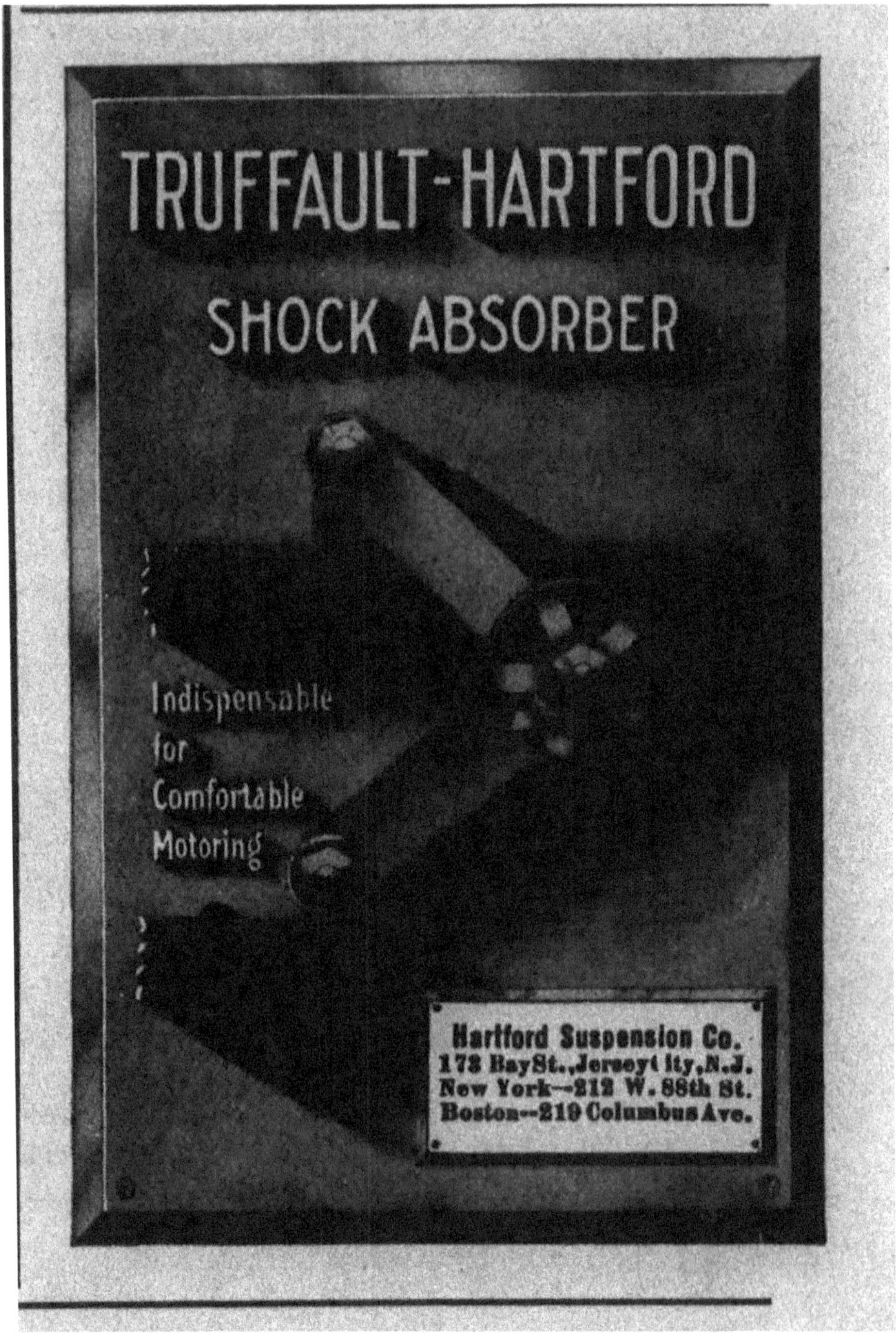

TRUFFAULT-HARTFORD
SHOCK ABSORBER

Indispensable
for
Comfortable
Motoring

Hartford Suspension Co.
172 Bay St., Jersey City, N.J.
New York—212 W. 86th St.
Boston—219 Columbus Ave.

THE
ACME
CARS

again demonstrated their superiority and reliability for all touring conditions in the L. I. A. C. Economy Run on February 25th.

Personal demonstration at any time by telephoning

J. W. MEARS

7-9 Ocean Parkway,   'Phone 1454 Flat.

1907 MATHESON

Licensed Under Selden Patent.

TOURING CARS, ROADSTERS, LIMOUSINES and LANDAULETS.

THE MATHESON COMPANY of New York, 1619-21-23 Broadway.   TEL. 4876 COL.

REPRESENTED IN

Chicago, Ill.—The Matheson Co. of N. Y., Western Branch, 1321 Michigan Ave.
Phila., Pa.—Titman, Leeds & Co., Broad & Cherry Sts.
Boston, Mass.—The Matheson Co., of Boston, 92 Mass. Ave.
St. Louis, Mo.—The South Side Auto Co., 2339 So. Grand Ave.
San Francisco, Cal.—The Matheson Co. of California, Cor. Golden Gate and Van Ness Aves.
Rochester, N. Y.—Rochester Automobile Co., 150-170 South Ave.
Albany, N. Y.—Albany Garage Co., 28-30 Howard St.
Binghamton, N. Y.—H. D. Clinton Auto Co.
Cleveland, O.—Central Auto & Supply Co.
Watertown, N. Y.—Watertown Auto & Supply Co.
Newark, N. J.—The Matheson Co. of New Jersey, 9 Clinton St.
Jersey City—Hudson County Auto Co.
Long Branch, N. J.—Long Branch Auto Co.
Baltimore, Md.—The Matheson Co. of Maryland—R. E. Wood Lumber Co.
Calgary, Alberta, Canada.—A. S. Urquhart.
Richmond, Va.—Motor Transfer Co.

The HEALY RIMS

(Demountable Type)

You have read about it. *Don't fail to see it* at the Show

Madison Square Garden Show, No. 501

Your car should be equipped now. Let us show you why.

Healy Leather Tire Co.
90 GOLD STREET - - **NEW YORK**

WINTON

CARLSON AUTOMOBILE CO.
1060 BEDFORD AVENUE.

A 30 H. P. car with OFFSET CYLINDERS is a better Purchase than a 40 H. P. car with Old-Style Cylinders.
This is a mechanical fact that we shall be glad to prove to you personally. Save a year's trouble and expense by investigating now.

Many Cars in One.

40 H. P. Latest Rothschild Limousine Body

The Crowning Achievement in French Automobile Construction

The entire mechanism is absolutely independent of the body, which may be removed with the greatest ease, to allow of a thorough inspection, or for the substitution of a body of different arrangement.

Delaunay Belleville

Has the strength of the World's Greatest Battleships

The largest Men of War in the English, French and Japanese Navies, also the leading factories in the United States are using the famous DELAUNAY BELLEVILLE BOILERS. The same superfine materials and workmanship give the remarkable strength for which these Automobiles are famous.

King Alfred, English Navy. 30,000 H. P.

American Branch, 1778 Broadway, New York.
H. NEUBAUER, Manager.   Tel. 6177 Columbus

"The Car Which Needs No Coaxing"

"The Maxwell" Automobile

For Substantial Every Day Use

I would like to prove to you that the "MAXWELL" is the one car that will give you continued and unqualified satisfaction. I can do it.

MODEL "H C"
16-20 H. P.—$1,450
Fully Equipped
TOURING CAR

MODEL "N C"
16-20 H. P.—$1,350
Fully Equipped
RUNABOUT

Two Cylinder.   90 Inch Wheel Base.   30x3½ Wheels.

"If you can't come to me, I'll come to you."

I. C. KIRKHAM
Exclusive Distributor, Long Island
1060 Bedford Avenue.

Tel. 4300/301 Bedford.

The Sensitive Indicating Arrow

of the JONES SPEEDOMETER responds to EVERY CHANGE OF SPEED. The tendency of many indicators is to "drag" from one speed to another, and seldom indicate accurately. This error was very much in evidence in the A. C. G. B. & I. speed indicating contest. The

JONES SPEEDOMETER

was the only instrument in the trials to score an absolutely

PERFECT RECORD

Jones Speedometer
154 W. 32d St
New York

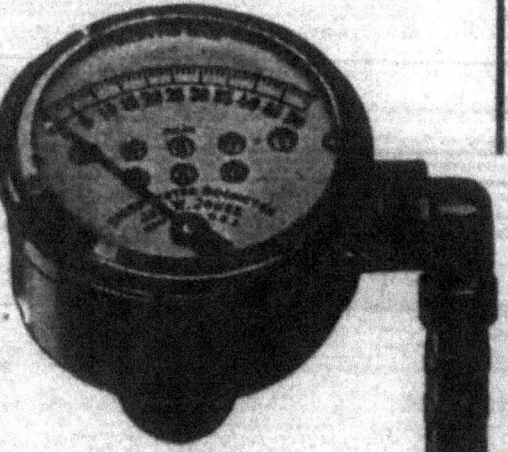

We exhibit at Garden Show, Jan. 12, 19.

LOOK AT THE CAR
—THEN—
LOOK AT ITS RECORD

The
OLDSMOBILE

This year you do not have to try a car, you get a tried car. The new Oldsmobile is the best car for 1906 bettered for 1907. Improvements but no experiments. Oldsmobile construction is standard.

Palace Touring Car and Flying Roadster $2,750.

Ask for our representative at the show

Brooklyn Motor Car Co.
1386 BEDFORD AVE.,
ROY E. PARDEE, Mgr.   Phone 241 Prospect

America's Most Famous Motor Car
The ROYAL TOURIST

See Our Exhibit at New York Show

MANY PEOPLE from all over the country have visited the enlarged factory of The Royal Motor Car Company at Cleveland to look over the new 1907 Royal, Model G, Series 2, car.

They have pronounced it the finest car in the world. They have been surprised to find an American factory producing a car which is very evidently destined to lead the whole automobile world this year.

It is built right, sold right and will run right. We hope you will take time to talk with some owner of a Royal Tourist.

The ROYAL MOTOR CAR CO.
CLEVELAND, OHIO

Member Licensed Association of Automobile Manufacturers

The Sign of Prestige

Charron, Girardot, Voigt

¶ The C. G. V. Car is so associated with the great names in the social and financial worlds of Europe and America, that it has become a hall-mark of prestige and social standing.

¶ There is nothing to compare with the C. G. V. in finish and in style. It is the *cheapest* car you can buy; it does not begin to deteriorate the first month of its use but remains for years a permanent value of its full purchase price.

¶ No matter what car you may buy, you will be enabled to get a greater value out of it by knowing about the C. G. V. Write for our superbly illustrated Catalogue No. 7, it will instruct you so thoroughly in the essentials of a good car that when you do go out to select one, no matter what make—or price—you will *know* the points of construction which make for power, durability, ease of control and elegance. Write to-day—*now*

Sole Importers for United States and Canada

C. G. V.
IMPORT
CO.

1849 Broadway
at 61st. St.
NEW YORK
CITY

35-40 H.P., $3,500.

POPE = TOLEDO

The most reliable, speediest and fastest hill climber of the year.

Best in design. Ample room for seven. Fully equipped.

Exhibited this evening at L. I. Auto Club.

Demonstration by appointment.

A. G. Southward,
342 Flatbush Ave., 10 Clinton St., Brooklyn.

Waverley ELECTRICS

FOUR-PASSENGER VICTORIA COUPE WITH REMOVABLE COUPE TOP

$2150.00

ESPECIALLY, IN WINTER

When large numbers of gasoline flyers are in storage, and horses demand more "humane" treatment than ever—

The Waverley Electric is the one vehicle that asks no favors from the weather man.

Winter, or summer, rain or shine, it's simply a case of "turn on the current and steer." Nothing to break down. No tinkering. No noise. No dirt. No odor.

The coupe top is removable, converting the car into an equally stylish Victoria Phaeton for summer use. This feature can be had in several Waverley models.

The Waverley Electric is the most dependable of all vehicles. It dispenses with the coachman. It needs no chauffeur.

No other vehicle gives such long and splendid service at so low a cost per mile.

Will Exhibit at Madison Square Garden Jan. 16-23.

THE POPE TOLEDO gasoline cars will be shown at Space 5 in the Garden.

A. W. BLANCHARD, Inc.
METROPOLITAN AGENT
342 Flatbush Ave., Bkyn.   1876 Broadway, N. Y. C.

Victoria Phaeton without coupe top $1700

Not at the Garden Show

The Complete Line of 1909

Buick

Models is on Exhibition in our spacious New York Salesroom.

Here you can examine at your leisure, away from noise and confusion, the latest models of the largest maker of automobiles in the world. Ask for a Brooklyn representative.

Open Evenings

BUICK MOTOR COMPANY
42 Flatbush Ave.

Telephone 4176 Main.
NEW YORK, 1733 Broadway (55th St.)
NEWARK, 222 Halsey St.

BAKER
ELECTRIC VEHICLES

QUEEN VICTORIA
Body Interchangeable with Extension Front Coupé

EXTENSION FRONT COUPÉ
Body Interchangeable with Victoria

MODEL R. RUNABOUT
Body Interchangeable with R. Coupé

MODEL R. COUPÉ
Body Interchangeable with R. Runabout

Baker Electrics lead all others in speed, efficiency and mileage.
Complete exhibit at Madison Square Garden Show, January 16th to 23rd.

BAKER VEHICLE CO.
Phone 2830 Columbus 1788 Broadway, cor. 58th St., N. Y.

The New STEVENS=DURYEA MODELS XXX and Y
are typical Stevens-Duryea productions

The latest links in the chain of STEVENS-DURYEA SUCCESSES.

The XXX—a 24 Horse Power Four Cylinder Runabout—Price $2850.

The Y—a 6-40 Horse Power Six Cylinder Touring Car—Price $4000.

The Four Cylinder Model X, of the past season, and Six Cylinder Model U (Light Six) of the past two seasons, will also be 1909 STEVENS-DURYEA CARS.

Now on Exhibition at 1909 Licensed Association Auto. Show at Madison Square Garden.

I. M. ALLEN CO
116 South Portland Ave., Brooklyn
Manufactured by Stevens-Duryea Co.
Chicopee Falls, Mass., U.S.A.
Members A. L. A. M.

All Cars Sold by Us Are Licensed Under Selden Patent and Guaranteed for One Year

EXHIBITING AT MADISON SQUARE GARDEN

THE Palmer-Singer line of cars—five types varying in horse power and price—was the motor car sensation of 1908. So immediate and so extensive was the demand in 1908 for our lightning-fast Six-Sixty, our speedy, snappy, graceful Skimabout, that we were unable to manufacture enough cars to supply purchasers in New York City alone. Now, with enlarged factory facilities, for the first time, we give the country at large an opportunity to learn of how much real motor car value may be purchased for a moderate sum. The policy of this house is and always has been to build cars which would be unquestionably the best in the world, of their type and class, entirely irrespective of price. No cars at any price can surpass the Palmer-Singer line in excellence of materials used, in quality of workmanship, in design, in style, in efficiency, performance and durability.

A Town Car suitable for Country Touring.

Palmer-Singer Town and Country Car—28-30 H. P.—$3500, is powerful enough to take its full complement of passengers on long, hard tours over any roads that any touring car can traverse. Open it is a light, comfortable Landaulet for country use, with an ever-ready shelter in case of storm. It is an extremely elegant summer car. Closed it is an ideal winter car—automatically heated, with the temperature always at the control of the passengers. For winter touring it has no equal. It far surpasses the heavy imported foreign town cars which from their heavy bodies and low power are entirely unfitted for country use. As a town car it is luxurious and beautiful, with every accessory to looks and comfort. It is the most completely serviceable car that money can buy. You should really know its details as our catalog shows them. It is very much worth while.

Sixty-five miles an hour guaranteed

Palmer-Singer Six-Sixty—6-cyl., 60 H. P., 65 miles an hour guaranteed—$3500. This is the Motor Sportsman's Car de Luxe. We guarantee 65 miles an hour. Neither in this country nor abroad is there anything of its type and class to compare with it. It gives power and speed in a degree that mere words cannot make you appreciate. The Six-Sixty is a 6-cylinder, 60 H.P. speed car capable of record-breaking speed and still of carrying from two to five passengers in luxurious comfort on long, hard runs. In durability and economy of upkeep it has, like our other cars, made a record which made us utterly unable to supply the demand for it in 1908 and which makes it the fastest selling car on the market to-day.

P. & S. Simplex

For quick delivery, ten 90 H. P. P. & S. Simplex Cars, this style. Savannah racing type— 90 miles an hour guaranteed. $6000—the ten fastest stock cars in America.

The Famous P. & S. Simplex—50 H.P., Krupp Steel throughout; built in New York City, 70 miles an hour—$5750. We are the sole distributors of the famous Simplex, $5750; we chase the entire output of the Simplex factory. The Simplex has long been celebrated as the best car made in America and without superior in the world. It is the holder of Sixteen world's records. It has repeatedly defeated the best cars of Europe and America in speed and endurance contests. A Simplex stock car won the great International 24-hour race at Brighton Beach, Oct. 2d and 3d, 1908, beating by the tremendous margin of 70 miles the best previous similar 24-hour performance of any car in the world and defeating by the same margin a large field of the best cars from this country and abroad. The same car, a stock car remember, defeated all American cars and four foreign makes entered at Savannah, although it was considered hopeless for a stock car to compete against specially built racers. The Simplex is built of Krupp steel throughout. It is standardized. When you buy a Simplex you KNOW that your car will TEN YEARS hence give you mile-a-minute speed, tremendous power, the maximum of durability and economy, and reliability. It gives you ten dollars' worth of automobile for every dollar you pay.

Specifications Common to All Palmer-Singer Models

Nickel steel is used to give lightness and strength. Imported F. & S. ball bearings exclusively. Bosch high-tension magnetos. Multiple disc clutches. Double and single drop frames. Drop forged I beam front axle—four-speed selective type, sliding gear transmissions, with direct drive on third speed. Brakes equalized, all expanding type and on rear wheels. Universal joints on all steering connections. Shaft-driven, all moving parts inclosed in dust-proof cases.

Palmer & Singer Mfg. Co.
1521 Michigan Avenue, CHICAGO
1620-22-24 Broadway, NEW YORK
Sole Distributors the Simplex

An American car, designed by an American, built by Americans in an American factory, of American material.
That's the

Locomobile

America's Representative Car

The I. S. REMSON M'f'g Co.
Brooklyn and Long Island Agents
740-750 Grand Street, Brooklyn, N. Y.
W. H. KOUWENHOVEN, Manager Auto Department

MOTOR CARS
"ASK THE MAN WHO OWNS ONE"

Special show-week display in addition to the Packard Exhibition at Madison Square Garden : :

Touring Cars
Limousines
Landaulets
Runabouts
Close-Coupled
and other special bodies.

Packard Motor Car Company
OF NEW YORK
1861 BROADWAY

New Models of Automobiles to be Shown at the Madison Square Garden Exhibition.

THE PIERCE-ARROW "SIX-FORTY-EIGHT", LANDAU.

THE 1909 PACKARD "THIRTY" LIMOUSINE.

THE NEW "LIGHT SIX" LOZIER.

THE KNOX MODEL O 38 H.P. TONNEAUETTE.

Ask for a Brooklyn Representative at the Garden Show

Cadillac

"Thirty"

$1400 Car Is Here

Call and inspect it or phone us for demonstration.

IT IS AMERICA'S GREATEST PRODUCTION

JOS. D. ROURK, 1001-3 Bedford Av.

DISTRIBUTER

PHONE 3730 BEDFORD.   OPEN EVENINGS.

HAYNES 1909 HAS ARRIVED
AS GOOD AS THE BEST--BETTER THAN THE BEST

MEET US AT THE SHOW

The Manufactured Not Assembled Car

A Car in the $5,000 class that Sells at $3,000

Borough Automobile Co.

BROOKLYN AND LONG ISLAND AGENTS

679-81-83 McDONOUGH STREET
BROOKLYN

TELEPHONE 3309 BUSHWICK
NIGHT CALLS 3130 BUSHWICK

AUGUST D. FINK, Proprietor

Auto Repairing and Overhauling

Painting and Trimming

See us at Madison Square Show, Jan. 12-19 at Winton Booth
Directly Opposite Main Entrance
Our Model M. Type X-I-V and Model M. Limousine demonstrators are here
CALL OR WRITE FOR DEMONSTRATION

THE CARLSON AUTOMOBILE CO.   1060 Bedford Avenue
Brooklyn

$33\tfrac{8}{10}$ Per Cent of All Cars in
THE GARDEN SHOW
(And $41\tfrac{3}{10}$ Per Cent of All American Cars) Are Equipped With

WRAPPED TREAD TIRES

And Every Diamond Tire Was Bought and Paid for by the Users.

74 Cars Have Diamond Wrapped Tread Tire Equipment, and
34 Cars or less than 16 per cent of the whole, are equipped with nearest competing make. **WHY?**

The Diamond Rubber Co., Akron, O.
1717 Broadway and 78 Reade St., N. Y.

1907 MATHESON
Licensed Under Selden Patent.

35 H. P. Touring Car - $4,500   50 H. P. Touring Car - $5,500

The Matheson Company of New York
1619-21-23 BROADWAY.

We exhibit at Madison Square Garden, January 12th to 19th, 1907.

1909 *Overland*   30 H. P.

H. T. Magneto on all Cars

Touring Car

$1,500

Three Speed Selective Sliding Gear

ROADSTER, $1,250.   THE SIMPLEST CAR TO CONTROL—ONLY PEDALS TO PUSH

Not one dissatisfied owner out of all the cars we have sold—150 owners in this section. Get a list and investigate.

OVERLAND CO. OF NEW YORK

1657 BROADWAY   -   Brooklyn Store: 1287 BEDFORD AVENUE

JEWEL CARS—THE HIT OF 1909
GUARANTEED FOR EVERY NIGHT AND DAY OF THE YEAR.

JEWEL 40 H. P. TOURING CAR $3,000
Complete with top, Hartford Shock Absorbers, Prest-O-Lite gas tank, 9" Rushmore Search lights, side lamps and tail lamp, double tire irons and trunk rack.
Wheel Base 120 inches. Transmission selective type. Three speeds forward and reverse. MOTOR 4¾ inch bore. 5-inch stroke.

AMPLE SPEED AND POWER FOR ANY EMERGENCY
Jewel Cars have always been deservedly famed for
HONESTY OF CONSTRUCTION AND DURABILITY
JEWEL MOTOR CAR SHOW AT OUR SALESROOMS
The Jewell Motor Car Co. of New York, Inc.
J. P. STOLTZ, President   1662 BROADWAY, NEW YORK CITY   Telephone, Col. 2618

40 H. P. Touring Car, $3,000   40. H. P. Limousine, $4,000

Demonstrations by appointment cheerfully given at any time.

THE RAUCH-LANG EXTENSION COUPE.

THE BOARDMAN

PRIVATE EXHIBITION

15 Imported Cars—Direct from Paris

You can't see them at Madison Square—but at Demarest Salesrooms—Fifth Avenue at 33rd Street, January 16th to 28th.

RENAULT

"The Car"
"Guaranteed for Life"

The most complete line ever handled by any manufacturer. Chassis built specially for *American roads*. Our motto—"Silence, Speed, Simplicity."

All types of Bodies—High Grade and Standard

RENAULT, "The Pride of France," leads the world in automobile construction. Is copied in every country. Builds 5000 cars a year. Employs 2800 workmen. Has delivered 6000 "Taxis" for Paris, London, Berlin, New York, Chicago and San Francisco.

1909 Prices:
8/10 H. P. 2 cylinder Runabout, $1750

9/12 H. P. 2 cyl. chassis	$2000	14/20 H. P. 4 cyl. chassis	$4000
10/14 H. P. 2 cyl. chassis	2500	20/30 H. P. 4 cyl. chassis	5000
10/14 H. P. 4 cyl. chassis	3000	35/45 H. P. 4 cyl. chassis	6000
12/14 H. P. 4 cyl. chassis	3250	50/60 H. P. 6 cyl. chassis	7500

Renault Freres Selling Branch
PAUL LACROIX, General Manager
Broadway and 52d Street, New York

Don't forget the address of this exhibition—"Demarest Show Rooms," Broadway and 33d Street. 15 new Renaults — new and classy French models—you can't see anything like them at Madison Square Garden.

The Kelly-Springfield Pneumatic Tires for Automobiles

The best known carriage tire in the world and the one that has been best known for the longest time is the Kelly-Springfield Tire. The Kelly-Springfield Pneumatic is made by the same manufacturers. It has the same wear-resisting, resilient composition.

CONSOLIDATED RUBBER TIRE COMPANY
20 Vesey Street, New York, and Akron, Ohio
Branches: New York, Boston, Chicago, St. Louis, Philadelphia, Detroit, Cincinnati and San Francisco

THE PIERCE-ARROW "SIX-FORTY-EIGHT", LANDAU.

THE NEW "LIGHT SIX" LOZIER.

THE 1909 PACKARD "THIRTY" LIMOUSINE.

THE KNOX MODEL O 38 H.P. TONNEAUETTE.

THE WHITE STEAMER IS THE ONLY CAR
of Distinctively American Design

The White is the only car which is not a copy or an imitation of some foreign product. In almost every class of machinery American ideas of construction have eventually proved triumphant, and so it is with the White Steamer. The White is sold in quantities abroad in competition with the home product, and, as regards this country, there are more Whites in use than any other make of large touring car.

The White possesses so many points of superiority over other types of automobiles that any one who purchases a car without first investigating the White is acting with only a partial understanding of the possibilities of automobile construction. We can meet the requirements of almost any pocket-book with either our 20 horse-power car at $2000 or our 40 horse-power at $4000 (shown above). The United States Government, the most discriminating of purchasers, owns more White Steamers than all other makes combined. Our cars are used by the War, Navy, and Executive Departments.

Write for descriptive matter

THE WHITE COMPANY
CLEVELAND, OHIO

NEW YORK CITY, Broadway at 62nd Street
BOSTON, 320 Newbury Street
PHILADELPHIA, 629-633 North Broad Street
PITTSBURG, 138-148 Beatty Street

CLEVELAND, 407 Rockwell Ave.
CHICAGO, 240 Michigan Ave.
SAN FRANCISCO, Market St. at Van Ness Ave.
ATLANTA, 120-122 Marietta Street

The sensational Roebling-Planche, 50 H.P. runabout, 4 cyl., bore 5½ in., stroke 5½ in.

Superior Design **MOON** Speed and
Absolute Simplicity   Economy of Operation

THESE ARE THE NOTEWORTHY FEATURES OF THE 1909 MOON

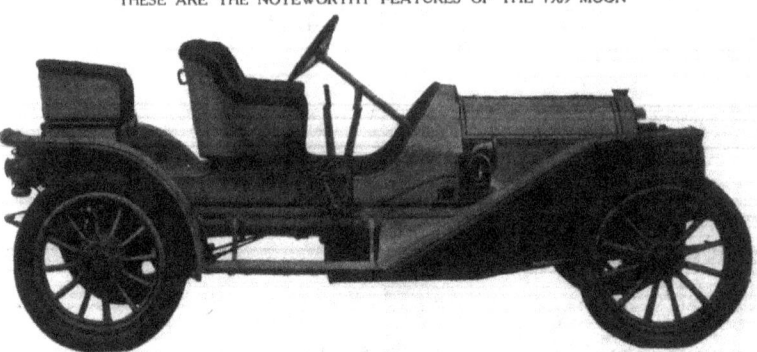

1909 MOON ROADSTER

¶Five years of consistent progress on one admirable model—the best made, better year after year—that is why the Moon Car is to the automobile buyer

THE BEST VALUE ON THE MARKET TO-DAY

¶Here are some of the points which make it so: Four speed and reverse selective type transmission, ball bearings, magneto ignition, multiple disc clutch, arched type of rear axle, 36 inch wheels, genuine honeycomb radiator, aluminum bodies, and above all, the most simple and efficient four-cylinder motor ever produced, furnished with Toy Tonneau Touring Car or enclosed bodies.

MOON MOTOR CAR COMPANY OF N. Y.
1761 BROADWAY, NEW YORK CITY

Telephone 6995 Columbus

THE INCOMPARABLE
WHITE
THE CAR FOR SERVICE

Features of the New White Steam Car

By the improved system of regulation in the new White cars, the steam pressure remains constant under all conditions. The person driving one of the new models for the first time will be able to get the same results as the most experienced operator. Added to this feature are the characteristic features of absolute silence, freedom from vibration, the absence of all delicate parts, genuine flexibility (all speeds from zero to maximum by throttle control alone), and supreme reliability.

The efficiency of the power plant has been so developed that the new models will run at least 150 miles on one filling of gasolene and water tanks.

The White Steam Car is now in its seventh year before the public. Its record from the first has been one of consistent success. We have built at least twice as many large touring cars as any other maker in the world, and therefore the purchaser of a White receives the benefit of an experience in designing and building not to be found elsewhere.

WRITE FOR DESCRIPTIVE MATTER

WHITE GARAGE

42 W. 62d St., New York.   1380 Bedford Ave., Brooklyn.

Chalmers–Detroit

No Other Car Like This For $1500.00

Never before have automobile builders included in the smaller type car all the features of the large and expensive types. Each point has been carefully worked out, and we surprise you with the completeness of detail and excellence of finish.

A large production by a factory with immense resources—not afraid to contract for material during the depression (of course at a tremendous saving) has made possible this car, and impossible any duplication by a competitor at anywhere near the price.

Over 1000 Models for 1909 already in the owners' hands convincingly demonstrate all our claims.

The Chalmers-Detroit Motor Co. have run, under close inspection, a $1500.00 Model 208 miles per day for 100 consecutive days—*20800 MILES*.
We have the good fortune to have this car in our salesroom. We would be pleased to have you ride 100 miles in it, anywhere, and decide for yourself upon the desirability of ownership, and the permanency of investment if you purchased a duplicate.

$2,750

In luxury and appointments you actually could not distinguish between this limousine and any $4,000 closed car sold in New York or Brooklyn

---

Carl H. Page & Company
Broadway at 50th Street - - - New York
Eastern Distributers
We exhibit only at Madison Square Garden AUTO. SHOW, January 16th to 23rd.

MR. KINGSLEY SWAN AND HIS NEW MODEL OF STEARNS CAR.
With him are Mr. Raymond Calvi and Mr. Francis Warren Kane.

"THE CAR WITH A ROAD-ESTABLISHED REPUTATION"

The "Maxwell" AUTOMOBILE

Seven Models for your Inspection at
THE PALACE SHOW

Ask for a BROOKLYN REPRESENTATIVE

I. C. KIRKHAM
EXCLUSIVE DISTRIBUTOR FOR LONG ISLAND
1060 BEDFORD AVE.

THE MODEL "M" WHITE STEAMER
is the Most Luxurious Car on the Market

The luxury of the White Steamer is unequalled by any other machine. There is no car, however expensive, which has better upholstery, better springs or better finish. Owing to the compactness of the mechanism, the body is much more commodious than in any car of similar wheel-base (122 inches). Few other makes have such a liberal tire equipment (36x4 on the front wheels and 36x5 on the rear wheels).

The luxury of any car, however, depends largely upon its riding qualities, and here the White is "in a class by itself." It is the only car where starting and changes of speed can be effected without jerks or jolts. It is the only car which runs noiselessly under all conditions. It is the only car which is free from vibration, because it is the only car where the power is applied evenly and continuously, and not spasmodically. It is the only car which does not, at times, emit malodorous vapors.

Because of its luxury and of its unequalled reliability the White has long been the favorite car with discriminating purchasers on both sides of the Atlantic. For example, the White has been used for several years by the gentleman who is most prominent in the industrial affairs of this country and by the gentleman who has been most prominent in the political affairs of this country.

Just as our Model "M," priced at $4,000, is the most desirable of the higher priced cars, so our Model "O," at $2,000, predominates among moderate priced machines.

Write for descriptive matter.

THE WHITE COMPANY
CLEVELAND, OHIO

NEW YORK CITY, Broadway and 62d Street
BOSTON, 320 Newbury Street
PHILADELPHIA, 629-33 North Broad Street
PITTSBURG, 139-148 Beatty Street

CLEVELAND, 407 Rockwell Avenue
CHICAGO, 240 Michigan Avenue
SAN FRANCISCO Market Street at Van Ness Avenue
ATLANTA, 120-122 Marietta Street

BAKER ELECTRIC VEHICLES.

QUEEN VICTORIA.

This Victoria Body is interchangeable with either Straight Front Inside Drive Coupe Body for two passengers or the Extension Front Inside Drive Coupe Body for four passengers.

The mileage and hill climbing ability of this car is greater than anything heretofore attempted in this class of vehicle.

Complete exhibit of new models at Madison Square Garden Show January 16-23. Also at our Showrooms

1788 Broadway   Cor, 58th Street
Phone 2830 Columbus.

Ninth National

AUTOMOBILE SHOW
MADISON SQUARE GARDEN
January 16 to 23, 1909
New York City

Under the auspices of the

Association of Licensed Automobile Manufacturers

Exhibiting standard Gasoline Cars licensed under the Selden Patent.

LICENSED GASOLINE CARS

Alden Sampson	Haynes	Pope-Toledo
Apperson	Hewitt	Royal Tourist
Autocar	Knox	Selden
Buick	Locomobile	Simplex
Cadillac	Lozier	Stearns
Chalmers-Detroit	Matheson	Stevens Duryea
Corbin	Oldsmobile	Studebaker
Columbia	Packard	Thomas
Elmore	Peerless	Thomas Detroit
E. M. F.	Pierce Arrow	Walter
Franklin	Pope-Hartford	WalthamOrient
	Winton	

ELECTRIC

Anderson	Baker	General Vehicle Co.
Babcock	Columbia	Rauch & Lang
Bailey	Columbus	Studebaker
Waverley		Woods

STEAM
White

Complete exhibit by the Motor & Accessory Manufacturers.
The only complete Motorcycle exhibit in New York by the Motorcycle Manufacturers' Association.
Commercial vehicles, Town Cars, Taxicabs, &c.

AUTOMOBILE
PAINTING and TRIMMING
Consult us for Estimate

THE I. S. REMSON M'F'G CO.
740-750 GRAND STREET   BROOKLYN

Pope-Toledo
and
Waverley Electric Automobiles

A. W. BLANCHARD, Inc.
342 Flatbush Av., 10 Clinton St.
Brooklyn, N. Y.
Telephone Prospect 3114

We carry a full line of **parts** for all **Pope** Cars.

A Good second hand **Hartford** for sale cheap.

35-40 H.P., $3,500.

POPE = TOLEDO

The most reliable, speediest and fastest hill climber of the year.

Best in design. Ample room for seven. Fully equipped.

Demonstration by appointment.

A. G. Southworth,

342 Flatbush Ave., 10 Clinton St., Brooklyn.

Overland

MAGNETO and FULL EQUIPMENT ON ALL CARS

OVERLAND CO. of N. Y.

We left Flatbush Monday morning and traveled around New York 28 miles. Left for Philadelphia, arriving in afternoon. Returned to New York Tuesday. Total mileage 276. Gasoline used, 14 gallons. No mechanical trouble. No tire trouble. Our car has run 10,000 miles this season, including one trip to Ohio. No quieter car can be found than it is to-day.—MRS. T. S. HILL. Jan. 24.

Roadster, 30 H. P., $1,250

Touring Car, 30 H.P., $1,500

THREE SPEED SLIDING GEAR
SELECTIVE TYPE

OVERLAND CO. OF NEW YORK
1657 BROADWAY

Brooklyn Store, 1287 Bedford Avenue

HAYNES 1909 HAS ARRIVED
AS GOOD AS THE BEST—BETTER THAN THE REST

The Manufactured Not Assembled Car

A Car in the $5,000 class that Sells at $3,000

Borough Automobile Co.
BROOKLYN AND LONG ISLAND AGENTS

679-81-83 McDONOUGH STREET
BROOKLYN

TELEPHONE 3309 BUSHWICK
NIGHT CALLS 3130 BUSHWICK

AUGUST D. FINK, Proprietor

Auto Repairing and Overhauling

Painting and Trimming

PRIVATE EXHIBITION

15 Imported Cars—Direct from Paris

You can't see them at Madison Square—but at Demarest Salesrooms—Fifth Avenue at 33rd Street, January 16th to 28th.

RENAULT

"The Car"
"Guaranteed for Life"

The most complete line ever handled by any manufacturer. Chassis built specially for *American roads*. Our motto—"Silence, Speed, Simplicity."

All types of Bodies—High Grade and Standard

RENAULT, "The Pride of France," leads the world in automobile construction. Is copied in every country. Builds 5000 cars a year. Employs 2800 workmen. Has delivered 6000 "Taxis" for Paris, London, Berlin, New York, Chicago and San Francisco.

1909 Prices:

8/10 H. P. 2 cylinder Runabout, $1750

8/12 H. P. 2 cyl. chassis $2000	14/20 H. P. 4 cyl. chassis $4000
10/14 H. P. 2 cyl. chassis 2500	20/30 H. P. 4 cyl. chassis 5000
10/14 H. P. 4 cyl. chassis 3000	35/45 H. P. 4 cyl. chassis 6000
12/14 H. P. 4 cyl. chassis 3250	50/60 H. P. 6 cyl. chassis 7500

Renault Freres Selling Branch
PAUL LACROIX, General Manager
Broadway and 57th Street, New York

Don't forget the address of this exhibition—"Demarest Show Rooms," Fifth Ave. and 33d Street. 15 new Renaults — new and classy French models—you can't see anything like them at Madison Square Garden.

BAKER
ELECTRIC VEHICLES

QUEEN VICTORIA
Body Interchangeable with Extension Front Coupé

EXTENSION FRONT COUPÉ
Body Interchangeable with Victoria

MODEL S. RUNABOUT
Body Interchangeable with R. Coupé

MODEL S. COUPÉ
Body Interchangeable with R. Runabout

Baker Electrics lead all others in speed, efficiency and mileage.
Complete exhibit at Madison Square Garden Show, January 16th to 23rd.

BAKER VEHICLE CO.

Phone 3830 Columbus   1788 Broadway, cor. 58th St., N. Y.

Announcement

Messrs. Hollander and Tangeman beg to announce that they have leased for ten years the modern fireproof building at the corner of 56th Street and Broadway, which will be the future American home of

FIAT
Automobiles

The building is being especially equipped for them, and they expect to take possession in the early part of January.

It is planned to incorporate a new Company to be known as

THE HOL-TAN CO.

which will succeed to the Hollander and Tangeman business.

This is merely a change of name, not of interests. The President of the Company will be Mr. C. H. Tangeman; the Vice-President and Treasurer, Mr. E. R. Hollander; and the Secretary, Mr. A. G. Hamilton.

Last year it was impossible for Hollander and Tangeman to meet the demand for "FIAT" Automobiles in this country. Arrangements have been made whereby deliveries during 1906 will be greatly increased, and it is the intention to have on hand at all times the various types of the famous "FIAT" cars, ready for the road when purchased.

Until removal, they invite inspection of "FIAT" cars at our old quarters.

3 and 5 West 45th St., N. Y.

| SOLE AMERICAN AGENTS | Licensed Importers under Selden Patent |

AGENCIES
HARRY FOSDICK CO., Boston, Mass.
ROCHESTER AUTOMOBILE CO., Rochester, N. Y.
DOMINION AUTOMOBILE CO., Toronto and Montreal, Canada
H. ALLEN DALLEY, Philadelphia, Pa.

"FIAT" cars will be exhibited only at Madison Square Garden Automobile Show

A. W. Blanchard, Inc.

MOTOR CARS

Toledo Gasoline. The best car built.
Ask the man who *knows*.

Waverley Electric. Simplest and best.
It gives more service at less expense per mile than any other vehicle.
Write for catalogue.

Everything for the motor car

342 Flatbush Av., 10 Clinton St., Brooklyn
1876 Broadway, New York

"Papa, why do brides wear long veils?" "To conceal their satisfaction, I presume, my son."—*Smart Set.*

AUTOMOBILE
PAINTING and TRIMMING
Consult us for Estimate
THE **I. S. REMSON** M'F'G CO
740-750 GRAND STREET   BROOKLYN

MR. C. H. TANGEMAN IN HIS HOL-TAN, THE FIRST ENTRY IN THE BRIARCLIFF RACE.

THE FAMOUS APPERSON "JACK RABBIT" FINISHING A RACE.

Ice, Mud and Roughly Frozen Roads serve only to emphasize the exclusive excellence of

Diamond
WRAPPED TREAD TIRES

Mr. R. G. Kelsey said

at the conclusion of his recent midwinter run in a Matheson stock car, from New York to Chicago:

"I will confess that at one time I was a staunch exponent of imported tires; but I am absolutely converted to Diamond Tires as being superior to anything with which I have ever had any of my machines equipped."

The Diamond Rubber Company
AKRON, OHIO

G. WARING STEBBINS

"A thorough education for the singer from the foundation up."
Authorized Teacher of the Sbriglia method.

One of Mr. Stebbins' most successful pupils is Mr. Wm. Howell Edwards, for eight years soloist Emmanuel Baptist Church.
Studios: 121 Carnegie Hall, New York
1171 Dean Street, Brooklyn
(Telephone: 665 J. Bedford)

Some girls, like canal boats, are not half as pretty as their names.—Puck.

Go To **Dieter's the Oldest Caterer of Brooklyn**
Parlor and Dining Rooms. Special arrangements for Receptions, Eucher Parties, Etc. Table d'hote 50 cts. 6 to 7 P. M. **913 Union Street**, near 8th Avenue. Trolley connects direct to house.

5 MONTAGUE TERRACE
overlooking Wall St. Ten minutes walk to Broadway, N. Y., neighborhood, house, table high-class.
Tel 905 Main

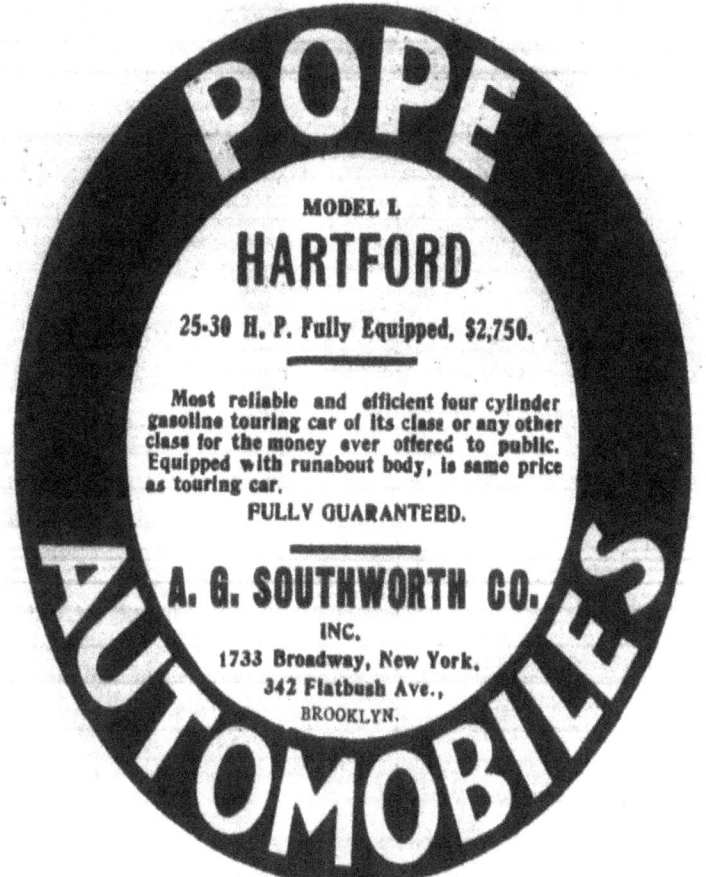

POPE AUTOMOBILES

MODEL L
HARTFORD

25-30 H. P. Fully Equipped, $2,750.

Most reliable and efficient four cylinder gasoline touring car of its class or any other class for the money ever offered to public. Equipped with runabout body, is same price as touring car.

FULLY GUARANTEED.

A. G. SOUTHWORTH CO.
INC.
1733 Broadway, New York.
342 Flatbush Ave.,
BROOKLYN.

Rambler
Leaders for 1906

Model
14
$1750

A medium weight touring car with four-cylinder vertical motor, 20—25 horse-power, sliding gear transmission, shaft drive and all modern appurtenances.
Wheel base, 106 inches; wheels, 32 inches, with 3½ inch tires.
Regular equipment includes: two gas headlights with separate generator, oil side and tail lights, horn, pump, tools, tire kit, etc.

Model
15
$2500

This car, somewhat larger and heavier than Model 14, is equipped with a four-cylinder vertical motor, 35—40 horse-power, giving one full horse-power to each 75 pounds dead weight of car fully equipped.
The power transmitting mechanism comprises a sliding gear set, with differential gear enclosed in the gear case, and individual chain drive.
Wheel base, 112 inches; wheels, 34 inches, with 4 inch tires.
Equipment same as Model 14.
Our catalog, giving full details of these and five other models, mailed upon request.

Main Office and Factory, Kenosha, Wis.
BRANCHES: CHICAGO, MILWAUKEE, BOSTON, PHILADELPHIA, SAN FRANCISCO.
New York Agency: 134 West 38th Street. Representatives in all leading cities.

THOS. B. JEFFERY & CO.

35-40 H.P., $3,500.

POPE = TOLEDO

The most reliable, speediest and fastest hill climber of the year.

Best in design. Ample room for seven. Fully equipped.

Demonstration by appointment.

A. G. Southworth,

342 Flatbush Ave., 10 Clinton St., Brooklyn.

"Pope Toledo"
MILE-A-MINUTE CAR

24 Horse Power - $3,500
14 Horse Power - $2,000

Noiseless and the realization of perfection in motor vehicles. Deliveries guaranteed.

WAVERLEY ELECTRICS

Our maintenance proposition will interest you. Pre-eminently the car for city use.

Exclusive Agent

A. G. SOUTHWORTH
342 Flatbush Ave.   10 Clinton St.
BROOKLYN

MAGNETO and FULL EQUIPMENT ON ALL CARS

OVERLAND CO. of N. Y.

We left Flatbush Monday morning and traveled around New York 28 miles. Left for Philadelphia, arriving in afternoon. Returned to New York Tuesday. Total mileage 276. Gasoline used, 14 gallons. No mechanical trouble. No tire trouble. Our car has run 10,000 miles this season, including one trip to Ohio. No quieter car can be found than it is to-day.—MRS. T. S. HILL. Jan. 24.

Roadster, 30 H. P., $1,250

Touring Car, 30 H. P., $1,500

THREE SPEED SLIDING GEAR
SELECTIVE TYPE

OVERLAND CO. OF NEW YORK
1657 BROADWAY
Brooklyn Store 1287 Bedford Avenue

BAKER
ELECTRIC VEHICLES

QUEEN VICTORIA
Body Interchangeable with Extension Front Coupé

EXTENSION FRONT COUPÉ
Body Interchangeable with Victoria

MODEL S. RUNABOUT
Body Interchangeable with S. Coupé

MODEL S. COUPÉ
Body Interchangeable with S. Runabout

Baker Electrics lead all others in speed, efficiency and mileage.
Complete exhibit at Madison Square Garden Show, January 16th to 23rd.

BAKER VEHICLE CO.
Phone 2830 Columbus   1788 Broadway, cor. 58th St., N. Y.

Not at the Shows

The Complete Line of 1909

Buick

Models is on Exhibition in our spacious New York Salesroom.

Here you can examine at your leisure, away from noise and confusion, the latest models of the largest maker of automobiles in the world. Ask for a Brooklyn representative.

Open Evenings

BUICK MOTOR COMPANY
42 Flatbush Ave.

Telephone 4176 Main.

NEW YORK, 1733 Broadway (55th St.)
NEWARK, 222 Halsey St.

The Kelly-Springfield Pneumatic Tires for Automobiles

The best known carriage tire in the world and the one that has been best known for the longest time is the Kelly-Springfield Tire. The Kelly-Springfield Pneumatic is made by the same manufacturers. It has the same wear-resisting, resilient composition.

CONSOLIDATED RUBBER TIRE COMPANY
20 Vesey Street, New York, and Akron, Ohio
Branches: New York, Boston, Chicago, St. Louis, Philadelphia, Detroit, Cincinnati and San Francisco

We are Ready

TO MAKE AN APPOINTMENT
WITH YOU

To Demonstrate

THE QUIET-MILE-A-MINUTE
1906--35-40 H. P.

Pope-Toledo

The highest type of advanced automobile construction. Strong, powerful, simple and exceedingly easy to control. Speed 4 miles to 60 miles on direct drive high speed

A. G. SOUTHWORTH

342 Flatbush Avenue   10 Clinton Street

STEVENS-DURYEA LIGHT SIX
Model U

It has been a source of gratification to the Stevens-Duryea Co. to be the recipients of a number of unsolicited testimonials regarding their Light Six Cars.

Because of the great pressure individual owners exerted to have them build an additional quantity of these models, they were obliged to materially increase their output for 1909. The Light Six Automobiles they will market considerably exceed in number the volume they had originally planned to construct.

The third consecutive year of the existence of the Stevens-Duryea Light Six Model U demonstrates how the popularity of and consequent widespread demand for this car is undiminished and unabated.

"Facts are Stubborn Things"

I. M. ALLEN CO
116 South Portland Ave., Brooklyn
Bet. Fulton and Hanson Place.
Tel. 4026 Prospect.

Manufactured by Stevens-Duryea Co.
Chicopee Falls, Mass.
Member A. L. A. M.

"POPE"

The Quiet Mile-a-Minute Car

14 H. P., $2,000
24 H. P., $3,500
40 H. P., $6,000

We can make early delivery of the new 24 H. P. 1905 Car.

Our orders for 1905 are placed; the demand for these excellent Cars is greater than ever. Orders for Spring delivery should be placed soon to insure early use of Car.

A. G. Southworth
342 Flatbush Ave., 10 Clinton St.
BROOKLYN

RACING EVERY WEEK-DAY

Until July 30. First Race at 2.30 P. M. Music by Mygrant's Band
SPARKLING PROGRAMMES OF STAKE, HANDICAP AND PURSE EVENTS, INCLUDING STEEPLECHASES

THE COOLEST PLACE AROUND NEW YORK

on a hot sultry day. The Grand Stand is open front and rear and on both sides, thus affording full sweep to the delightfully cool and bracing Ocean Breezes. Unobstructed view of Races from all parts of the Grand Stand and Club House.

BRIGHTON RACES

SENSATIONAL CONTESTS ECLIPSING THE WORLD'S RECORDS
ALL THE BEST HORSES OF THE SEASON IN SUPERB FORM

Saturday, July 23
The $10,000 BRIGHTON DERBY

together with the $7,500 Venus Stakes, and the Great Brighton Steeplechase over the full course
AND THREE OTHER STAR EVENTS

Brighton Course can be reached by Special Electric Trains on Brighton Road, and by Smith Street Trolley Cars. Special trains via Long Island Railroad leave Long Island City at regular intervals. Special Electric Trains leave 39th Street Ferry every 20 minutes.

WINTON

We can arrange now to give you a desirable delivery date. Later on that will not be possible.

CARLSON AUTOMOBILE CO.
1060 Bedford Avenue, Brooklyn

Announcement

Messrs. Hollander and Tangeman beg to announce that they have leased for ten years the modern fireproof building at the corner of 56th Street and Broadway, which will be the future American home of

FIAT

Automobiles

The building is being especially equipped for them, and they expect to take possession in the early part of January.

It is planned to incorporate a new Company to be known as

THE HOL-TAN CO.

which will succeed to the Hollander and Tangeman business.

This is merely a change of name, not of interests. The President of the Company will be Mr. C. H. Tangeman; the Vice-President and Treasurer, Mr. E. R. Hollander; and the Secretary, Mr. A. G. Hamilton.

Last year it was impossible for Hollander and Tangeman to meet the demand for "FIAT" Automobiles in this country. Arrangements have been made whereby deliveries during 1906 will be greatly increased, and it is the intention to have on hand at all times the various types of the famous "FIAT" cars, ready for the road when purchased.

Until removal, they invite inspection of "FIAT" cars at our old quarters.

3 and 5 West 45th St., N. Y.

Announcement

Messrs. Hollander and Tangeman beg to announce that they have leased for ten years the modern fireproof building at the corner of 56th Street and Broadway, which will be the future American home of

FIAT

Automobiles

The building is being especially equipped for them, and they expect to take possession in the early part of January.

It is planned to incorporate a new Company to be known as

THE HOL-TAN CO.

which will succeed to the Hollander and Tangeman business.

This is merely a change of name, not of interests. The President of the Company will be Mr. C. H. Tangeman; the Vice-President and Treasurer, Mr. E. R. Hollander; and the Secretary, Mr. A. G. Hamilton.

Last year it was impossible for Hollander and Tangeman to meet the demand for "FIAT" Automobiles in this country. Arrangements have been made whereby deliveries during 1906 will be greatly increased, and it is the intention to have on hand at all times the various types of the famous "FIAT" cars, ready for the road when purchased.

Until removal, they invite inspection of "FIAT" cars at our old quarters.

3 and 5 West 45th St., N. Y.

SOLE AMERICAN AGENTS   Licensed Importers under Selden Patent

AGENCIES
HARRY FOSDICK CO., Boston, Mass.
ROCHESTER AUTOMOBILE CO., Rochester, N. Y.
DOMINION AUTOMOBILE CO., Toronto and Montreal, Canada
H. ALLEN DALLEY, Philadelphia, Pa.

"FIAT" cars will be exhibited only at Madison Square Garden Automobile Show

BAKER
ELECTRIC VEHICLES
1909 MODEL

**INSIDE DRIVE
EXTENSION FRONT COUPÉ**

The New 1909 Model Four Passenger Inside Drive Coupé. This Coupe Body is interchangeable with Queen Victoria Body. Mileage and speed greater than heretofore attempted in this type of vehicle.

The Model "R" Runabout is another new 1909 model which will run over one hundred miles on one charge of battery. The Runabout Body is interchangeable with an Enclosed Body.

Complete Exhibit of the New Models at Madison Square Garden Show, January 16th to 23rd. Also at our Showrooms.

BAKER VEHICLE CO.

Phone 2830 Columbus. 1788 Broadway, cor. 58th St., N.Y.

The Looker-On at Coney Island.

LAST summer a magazine writer thus delivered himself of some thoughts that seemed to him to be fairly clamoring for expression: "Two generations ago Coney Island was a windswept waste of sand, stretched along the ocean's edge east of the opening of New York Harbor. A generation ago the waste was dotted with booths and hurdy-gurdies and bathing-houses. The island was a resort to which adventurous dwellers in Brooklyn journeyed at great expense of family, time and treasure, for a day's outing by the sea. Very few people in New York knew aught of it." This is picturesque enough, and sometimes magazines have been known to go in for picturesqueness at the expense of truth; but what rank nonsense such a statement is. The old Ocean Hotel, which still stands just west of the much larger hostelry at Brighton Beach, was built by W. A. Engeman in 1868. Then there was the old Vanderveer place in the heart of the west end and all in all Coney Island had progressed considerably beyond a waste "dotted with booths and hurdy-gurdies and bathing-houses" a generation ago—assuming that period to be a third of a century.

AS a matter of fact it is only four years short of a generation since the railroad to Brighton Beach was opened and at that time the west end was a pretty thriving place. The same year—1878—the United States Government Building of the Centennial Exposition of two years before was removed to Coney Island in sections and set up as the still existing Sea Beach Palace. The old Culver railroad was running more than a generation ago and prior to that there was the Gunther line, and horse cars as well, while for years before the middle seventies driving to Coney Island from Brooklyn was tolerably common, the old Pavilion Hotel at Far Rockaway attracting those who wanted to make their fast horses travel farther. Coney Island has really been accessible for about sixty years, or almost two generations, the first steam road and the horse car line being the eventual outcome of the erection of the Pavilion by Eddy and Hart, two New York speculators, in 1844. The Pavilion stood on the extreme western end of the island. More buildings were put up and the place so soon became famous as Norton's Point, that some sort of transportation facilities simply had to be considered. To say that "very few people in New York knew aught" of Coney Island only thirty-three and a third years ago is absurd. The writer was there himself in 1865 or 1866 and he then lived more than a hundred miles from New York. He has been told of the trip—which was taken from Brooklyn—again and again but never heard that it was "adventurous" or that it was undertaken "at great expense of family, time and treasure." If Brooklyn people had visitors then the proper thing to do was to take them down to Coney Island and New Yorkers, despite the longer journey, were somewhat given to the same happy custom—which fortunately has not died out in these days when Brooklynites are all New Yorkers.

I WAS reminded of the old isle of the conies—accepting for the moment this derivation of the name—by the electric sign which greets you on the outside of Luna Park this season. This sign reads, at night "The New Coney Island" and certainly Thompson and Dundy have a good enough right to display it. By the creation of Steeplechase Park, Mr. George W. Tilyou—whose "Walk" was long before one of the sights of this biggest suburban resort that any city in the world can boast of—took the first prodigious stride towards the transformation of Coney Island; but, after all, it was the Luna Park of Thompson and Dundy that, only a few years ago, literally made things hum and caused the universe to sit up and take notice. They builded better than they knew; for even such a dream or limitless dreams as Mr. Frederic Thompson, whom death soon ordained to shoulder their big amusement problem alone, could not, in 1902, have foreseen his permanent fairyland of to-day—with Dreamland, Happyland, Golden City and no end of lesser "parks" following in its wake.

THOSE who recall the earliest days of Luna Park—when its effulgent glory came as a complete revelation, even to those who had been electrically satiated at the Columbian and Pan American Expositions—have rather regretted the filling in of the vast central space a few seasons ago. There was a fascination in the very spaciousness, especially at the nearer end, where the crowd mostly congregated. Still there is more of a blaze at night and the free open air circus is a great attraction for the multitude. Then again the change brought into being those delightful balcony boxes, where, if one is able to secure places, one may dine al fresco, while the day fades, only to be brought back by the flooding light of myriads of incandescent lamps directly the electricity is turned on for the evening. Dining in those boxes is absolutely unique and this feature has been one of the chief means of bringing to Luna Park the contingent that does not buy pleasure with pennies, dimes, nickels and quarters, but on the contrary, thinks nothing of spending from twenty dollars upward. Moreover the neighboring Japanese Tea Gardens, also high above the street level, offer an exceptional opportunity for five-o'clocking, as the French have been wont to say since they began to make wholesale draughts on the English language. The Tea Gardens continue to have the highly picturesque mountain torrent chutes directly opposite—one of the new features of last summer—so that one may sip tea in Japan at five o'clock and divide his glances most agreeably between Nippon and a bit of Alpine scenery. There are anachronisms, to be sure; but then one does not need to analyze to the last point the little pleasures of a warm summer's day.

POSSIBLY as the last step in the way of the recent divorce from the Hippodrome, certain elephantine architectural adornments have disappeared from Luna Park. Each of the four corners of the base of the great tower of light now sports a female—airily gowned in a few yards of cloth and comfortably seated on a crescent moon. Elsewhere are more of the ladies of the new moon. The most amusing novelty of the season at Luna Park is unquestionably "The Tickler," the latest development of the bumpety-bump craze. The way the tubful of assorted humanity wabbles about on its highly erratic downward course makes one think of what must have been the early experiences of the three wise men who "went to sea in a bowl." One may also go up differently this season, as well as down, there being a new escalator on which you sit as on a horse. The men straddle the seat and the women take their choice between this and the other way. "The Wreck of the Corsair," which purports to offer "the greatest scenes on the greatest stage in the world"—Mr. Thompson does not have to bar the Hippodrome now; "The Days of '49," "Night and Morning" and the Brownie Theater are the new shows of importance. Out of doors there is a new rooster and ostrich merry-go-round, called "The Rounders," and also a gigantic sports bulletin. "The Old Mill," has become "The Red Mill"; the two monsters that guard the entrance to "The Dragon's Gorge" have turned blue—very blue; in a few other places the paint brush has had carte blanc with the paint pot; the flowers, that have ever been abundant here, are doing their best to make up for lost time, and—well, I guess that will be all just now.

OVER at Dreamland the paint bill has been infinitely heavier. Paint, in fact, has not done a thing but make the familiar seem unfamiliar there. Where once were vistas of immaculate whiteness there is now a riot of color, only the Bostock building having escaped. The tower is dark green at each of the four corners, as high up as the cornice, and on a bed of this color stand out orange and red eagles. But this is not a circumstance to "The Orient," which replaces the lilliputian village. Here there is color galore and much in the way of sculptural adornment that is very effective. Four camels and a donkey with native drivers are stationed before the entrance for the benefit of those who dote on that sort of thing. "The End of the World," next door, is quite as showy while farther along "Hell Gate" is not only redder and greener than ever but four red devils, of the most approved theatrical hue, are fresh bits of color. Across the way Roltair's "Arabian Nights" flaunts a giantess, by way of columns, for each day of the week and vies in bright color effects with the neighboring new shows, "The Great Divide" and "Yellowstone Park." What was once "Venice," has become "The Bay of Naples," and here instead of paint, a profusion of wistaria and purple clematis has been used to conceal the white facade. "The Alps" have also come in for more color, though only green was handed out to them. They drew very little of that by comparison with the open space between the tower and the Bostock building, where the open air vaudeville stage and cafe and the Maxim airships are. This space has been marked off by connecting arches of heavy greenery, of the same sort of prepared foliage that has adorned the entrance to the Herald Square Theater since "The Road to Yesterday" had its premiere there. The effect is extraordinarily

The Looker-On—Continued.

good, refreshing by day and still more alluring by night. The arches are high and broad; so there is just enough seclusion. In short, Dreamland, which began by being all art, is now all color. The public seemed to hanker after color and so it has got it.

* * *

AT Steeplechase Park the rose-gardens have been carrying the glories of June over into July, thanks to the late season. The gardens are not large but they make a fine showing and contribute color to a resort that has the advantage of plenty of trees. This year there are also trees where you start riding the horses—one of the very best attractions that Coney Island has ever had—but they are painted trees. Much of the fun centers about the "Human Roulette," near the entrance. Last summer there was only a small revolving table but this season there is a huge one, and from its edge the floor slopes sharply to a level space, with padded sides. It is a great place for "kids"—some of whom are smart enough to wear rubber-soled shoes and are thus able to withstand the centrifugal force of this clever amusement device. In the big ball-room roller-skating has so far encroached upon dancing that the latter has only half the usual space. The change is for the better, in more ways than one and certainly the bad skating is quite as amusing to the onlookers as the bad dancing. Not that all of either is bad; some of both is uncommonly good. At least once a summer and generally oftener, I drop into the French liner down at the other end of the park to see the panorama of the Mediterranean voyage. This panorama, which has been there for years, is one of the most beautiful illusions that I have ever seen, far excelling in realism most stage effects. The coloring, the perspective and the semblance of motion are all marvelous. This season the Friede Globe Tower has made quite a hole in Steeplechase Park on the Surf avenue side. The concrete foundations give some idea of the immensity of the thing, these piers extending over a huge space. Part of the steel structural work is on the ground and if all goes well, the biggest thing yet at Coney Island may be astonishing the blase and the unsophisticated hoi polloi alike a year hence. ADDISON STEELE.

MATHESON
The Standard American Car

SOME big beautiful foreign cars sell for $10,000 to $14,000. Are they worth it? Certainly they are. So is the Matheson; it is just as big; just as beautiful, and its performances over the roughest American roads, without a single repair or adjustment, far surpass the performances of any foreign car. For use in America the Matheson is worth **more** than any foreign car you can buy.

Reasonable Deliveries

Touring Cars
Runabouts
35 H. P. $4,500
50 H. P. $5,500

Guaranteed for One Year

Licensed Under Selden Patent

The Matheson Company
OF NEW YORK

Telephone 4875 Col. 1619-21-23 Broadway

Until motors abolish the horse we shall continue to make liveries for coachmen and grooms, as well as chauffeurs' outfits.

Illustrated catalogue on request.

ROGERS, PEET & CO.,

258-842-1260 Broadway,

(3 Stores)

NEW YORK

TAKE IT WITH YOU FOR THE OUTING

JUST OUT

THE SCARLET CAR

BY
Richard Harding Davis
The Best of Automobile Stories
Illustrated. $1.25

CHARLES SCRIBNER'S SONS

More than fifty of these popular cars are in use in Brooklyn. Ask their owners as to there qualities. The smoothest running, most economical touring car sold. We have in stock for immediate delivery the last shipments we will have this year of the

14 H. P. Car - $2,000

AT EITHER STORE

A. G. Southworth
**342 Flatbush Ave., 10 Clinton St.
BROOKLYN**

SUMMER MEETING

BEGINS WEDNESDAY, JULY 6

AND CONTINUES EVERY WEEK DAY UNTIL JULY 30

STAKE, HANDICAP AND PURSE EVENTS, INCLUDING STEEPLECHASES

MUSIC BY MYGRANT'S MILITARY BAND.

THE COOLEST AND MOST DELIGHTFUL PLACE AROUND NEW YORK TO GO FOR AN AFTERNOON'S DIVERSION

BRIGHTON RACES

SATURDAY, JULY 9

BRIGHTON HANDICAP

VALUE $25,000

Unobstructed View of Races from all Parts of the Grand Stand and Club House

Brighton Course can be reached from New York side of Brooklyn Bridge by Special Electric Trains on Brighton Road, and by Smith Street Trolley Cars. Special trains via Long Island Railroad leave Long Island City. Also via 39th Street Ferry, from foot of Whitehall Street, boats leave every 20 minutes, connecting with special Trolley cars.

MATHESON
The Standard American Car

A CAR that can carry seven people seventy miles an hour; that can speed over hundreds of miles of the roughest roads without the need of a single adjustment or repair is beyond all question the greatest car in the world. The Matheson has DONE this easily, not once but repeatedly.

Runabouts, Touring Cars, Limousines and Landaulets
Licensed Under Selden Patent

REPRESENTED IN

Chicago, Ill.—The Matheson Co. of N. Y., Western Branch, 1318 Michigan Ave.
Phila., Pa.—Titman, Leeds & Co., Broad and Cherry Sts.
Boston, Mass.—The Matheson Co., of Boston, 92 Mass. Ave.
St. Louis, Mo.—The South Side Auto Co., 2339 So. Grand Ave.
San Francisco, Cal.—The Matheson Co. of California, Cor. Golden Gate and Van Ness Aves.
Rochester, N. Y.—Rochester Automobile Co., 150-170 South Ave.
Albany, N. Y.—Albany Garage Co., 28-30 Howard St.
Binghamton, N. Y.—H. D. Clinton Auto Co.
Cleveland, O.—Central Auto & Supply Co.
Watertown, N. Y.—Watertown Auto & Supply Co
Newark, N. J.—The Matheson Co. of New Jersey, 9 Clinton St.
Jersey City—Hudson County Auto Co.
Long Branch, N. J.—Long Branch Auto Co.
Baltimore, Md.—The Matheson Co. of Maryland—R. E. Wood Lumber Co.
Calgary, Alberta, Canada.—A. S. Urquhart.
Richmond, Va.—Motor Transfer Co.

The Matheson Company of New York

Telephone 4876 Col.   1619-21-23 Broadway

Made for men who know correct livery by men who know good clothes.

Livery for every man-servant

Illustrated price list on request.

Special garments for-summer motoring.

Owner and chauffeur.

ROGERS, PEET & CO

258—842—1260 Broadway

(3 Stores)

NEW YORK

Flandrau & Co.

406, 408, 410, 412 Broome St.
Just off Broadway   Dozen Blocks from Bridge
New York City

WE show this Spring an unusually varied and interesting assortment of Carriages of every description for
TOWN AND COUNTRY

STATION WAGON

WE have for many years made a specialty of Carriages for Suburban places as well as for Town, and our productions can be depended upon to not only render long satisfactory service, but will ride right, will carry, and the Lamps and other fittings will be found practical in use.

AUTOMOBILE BODIES

WE build and finish these Bodies in all styles for any type of Chassis. Automobile painting and repair work of every description done on the premises. For nearly ten years we have been building Electric Carriages complete, and have turned out some of the most successful and handsomest Electrics made. Estimates furnished.

The "Maxwell"

Simplicity. Reliability.

An ideal car for the family.

Economy of operation.

EASY TO OPERATE

EVERY PART ACCESSIBLE.

The "MAXWELL" 20 H. P. Touring Car $1450, complete.

The above is a cut of the famous "Old 41" 1905 MAXWELL Touring Car, which came through the Decoration Day Endurance Run of the Long Island Automobile Club with a **perfect score,** covering 295 miles without raising the bonnet from the engine. This same car had previously completed two Glidden Tours with a perfect score, and in all has been driven over 20,000 miles.

Points that have made the MAXWELL famous:

1. Natural water circulation. No pump.
2. Three point suspension.
3. Engine and transmission case one aluminum casting.
4. All metal multiple disc clutch.
5. Shaft drive. No chains
6. All metal bodies.

WRITE FOR CATALOGUE

SALESROOM:
Bedford Ave. & Fulton Street, Brooklyn
TELEPHONE: 175 BEDFORD

I. C. KIRKHAM
Exclusive Distributor for Brooklyn and Long Island
Open Evenings

Garage:
399 SUMNER AVENUE

Healy Detachable Tire Grip

Prevents Skidding

Four times the wear of any other grip.

Easily Attached

Does not injure the tires

Makes no noise while running

Cheapest in the end

HEALY LEATHER TIRE CO.

88-90 GOLD STREET
Tel. 4794 Beekman

1906 BROADWAY
4487 Columbus

Thirty Horse Power

on the road is better than fifty in the catalogue. It is on the road you need it, and where, in the Model "H"

KNOX WATERLESS

PRICE, $2500,

you get it. It is why the "H" has outrun and outclimbed every other car in its class. Knox quality—standard for seven years—guarantees its wearing ability. Weight, 2250 pounds, actual; 102 inch wheelbase; 32 x 4 inch tires; three-speed selective transmission; positive automatic oiling; compensating carbureter; deadstop brakes; direct cooling; three-point suspension; straight shaft drive; long, easy springs; speed from 4 to 40 miles an hour. Write for illustrated catalogue and technical description; they are text books on automobile construction.

Knox Automobile Company

Member Association Licensed Automobile Manufacturers

SPRINGFIELD, MASS.

Brooklyn Representative, A. R. TOWNSEND, 1148 Bedford Avenue.

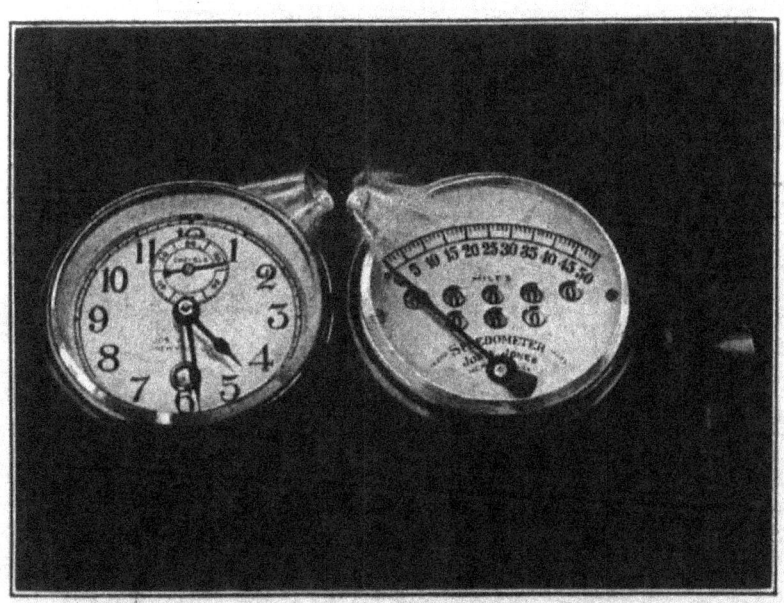

THE
"Day and Night" Jones

places at the instant command of the Motorist, Day or Night, Time—Speed—Total and Trip Mileage.

It is a combination of the Jones Auto Clock, and the Jones Speedometer-Odometer mounted together with small elecric light.

It embodies the dignity and tone characteristic of the Jones instruments and the incontestable accuracy guaranteed by the reputation of such.

We suggest that you write for The Speedomotor, a magazine on speed topics. You will find it worth many times its price. Its price is a request to Dept. 11.

JONES SPEEDOMETER CO.
2228 Broadway, New York

The Incomparable
WHITE
THE CAR FOR SERVICE

Fastest Time in Wilkes-Barre Hill-Climb

In the Wilkes-Barre Hill-Climb, held on May 30th, a stripped 30 horse-power Model "G" White Steamer made the climb up "Giants' Despair" in 1:49 4-5. This time is ten seconds faster than the best record made by a gasolene car. Practically all of the more prominent makes of cars were represented, there being over 50 contestants in this classic event.

Our record was made in a trial against time. We could not compete against anything else because the Contest Committee of the Wilkes-Barre Automobile Club barred us from all the regularly scheduled events, including the "free-for-all" because, as the President of the Wilkes-Barre Automobile Club informed us, in a formal letter returning our entries, "It was found that a large number of gasolene entries could not be received were steam cars allowed to compete in the gasolene events."

THE WHITE COMPANY
1380 BEDFORD AVENUE, BROOKLYN

ANTI-SKID
Diamond
ANTI-SKID TIRES
(Wrapped Tread)

American and Millimeter Sizes.
Quick, Detachable and Regular Clincher.
Most Efficient, Most Serviceable.
Hold on Any Surface and Stand the Strain.

THE DIAMOMD RUBBER CO., - Akron, Ohio.
New York - 1876 BROADWAY, - New York.

Racing Every Week-Day

AT 2:30 P. M.—UNTIL JULY 30.

Stake, Handicap and Purse Events, including Steeplechases.

Music by Mygrant's Band.

BRIGHTON RACES

Coolest and pleasantest place around New York to go for an afternoon's diversion

Unobstructed View of Races from all parts of the Grand Stand and Clubhouse

STAKE EVENTS.

Monday, July 11, Distaff Stakes.
Tuesday, July 12, Seagate Stakes.
Wednesday, July 13, Jamaica Stakes.
Thursday, July 14, Winged Foot Handicap.

Monday, July 18, Nautilus Stakes.
Tuesday, July 19, Sunshine Stakes.
Wednesday, July 20, Glen Cove Handicap.
Thursday, July 21, July Stakes.

Monday, July 25, Sea Gull Stakes.
Tuesday, July 26, Seashore Handicap.
Wednesday, July 27, Holiday Stakes.

Thursday, July 28, { Iroquois Stakes. Aintree Steeplechase.

Saturday, July 30, { Neptune Stakes. Brighton Oaks.

Saturday, July 16, { Curragh Steeplechase. Iolip Handicap. Brighton Junior Stakes.

Saturday, July 23, { Brighton Steeplechase. Venus Stakes. Brighton Derby.

All Routes to Coney Island lead direct to Brighton Course

All the crack horses of the year Fastest track in the world

MR. PHILIP HINES DRIVING A POPE-HARTFORD; MR. WINTHROP E. SCARRITT IS WITH HIM.

ROSE STAHL, NOW APPEARING AT THE HACKETT, IN HER NEW 26 H. P. DRAGON TOURING CAR.

Brighton Races

Racing within the sound of the ocean waves
MUSIC BY SUPERB MILITARY BAND
THE COOLEST SPOT NEAR NEW YORK

Star Events for Week July 15th to 20th

Monday—Atlantic Stakes for 2-year-olds
Tuesday—Sea Gate Stakes for 3-year-olds
Wednesday—Venus Stakes $7,500, for 2-year-old fillies
Thursday—Queen Stakes $7,500, for 3-year-olds
Saturday—Neptune Stakes, Curragh Steeplechase and Islip Handicap

All the best horses in training will meet in the above events

All Roads Lead to Brighton
SPECIAL ENTRANCE FOR CARRIAGES AND AUTOS NEPTUNE AVENUE

NEWPORT ESTABLISHMENT OPENED JUNE 1
Address 100 BELLEVUE AVE

BAKER ELECTRIC VEHICLES

Vehicles in our Newport Rental Service are stylish and up-to-date. The equal of any in private use.

Have stood the test of time. They represent the highest standard of Electric Vehicle Construction

To meet the exacting conditions of the twentieth century demand, six absolutely new models have been turned out that cover the entire range of the electric in the pleasure field. Six Superb Models that mark a new era in electric vehicle construction. Baker Electric Vehicles have always been noted for their great mileage capacity—the 1907 models are masterpieces in this respect as in all others. They surpass any mileage of any electric automobile made in this country or Europe—either on test for mileage or for speed.

Immediate Deliveries

ALSO

Landaulets, Broughams, Coupes,
Runabouts, Stanhopes, Suburbans,
Depot Carriages, Imperials, etc.

NEW BAKER RUNABOUT
A smart, novel and serviceable all around car for pleasure, riding or business use; strong, speedy, powerful. It has a greater mileage in any of its various speeds than any other electric car can demonstrate, with an equipment of regular stock batteries and tires.

C. B. RICE, 1790 BROADWAY, N. Y.
Tel. 2830 Col.

BAKER QUEEN VICTORIA
Rich and dignified in appearance, this is the most perfect vehicle for town use that has ever been conceived. It is distinctively smart and elegant in every detail.

Last Week of Summer Meeting

RACING ENDS JULY 30

SPARKLING PROGRAMMES OF STAKE, HANDICAP AND PURSE EVENTS, INCLUDING STEEPLECHASES

First Race at 2.30 P. M.   Music by Mygrant's Band

THE COOLEST PLACE AROUND NEW YORK

on a hot sultry day. The Grand Stand is open front and rear and on both sides, thus affording full sweep to the delightfully cool and bracing Ocean Breezes. Unobstructed view of Races from all parts of the Grand Stand and Club House.

BRIGHTON RACES

Particularly Attractive Stake Events to be decided the closing week

MONDAY, JULY 25—THE SEA GULL STAKES   TUESDAY, JULY 26—SEASHORE HANDICAP
WEDNESDAY, JULY 27—HOLIDAY STAKES   THURSDAY, JULY 28—IROQUOIS STAKES

Last Day—Saturday, July 30

The $7,500 Neptune Stakes   The $5,000 Brighton Oaks   The Aintree Steeplechase

TOGETHER WITH THREE OTHER STAR EVENTS

Brighton Course can be reached by Special Electric Trains on Brighton Road, and by Smith Street Trolley Cars. Special trains via Long Island Railroad leave Long Island City at regular intervals. Special Electric Trains every 20 minutes from 39th Street Ferry.

WINTON

CARLSON AUTO CO., 1060 Bedford Ave.

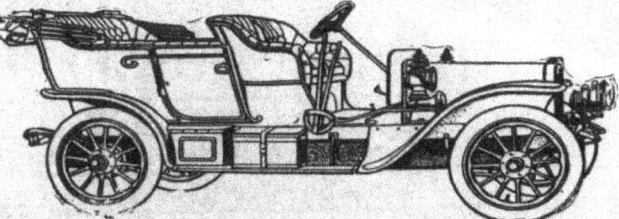

Perfect score in L. I. A. C. Endurance Test, May 30-31 and June 10

Winton cars are so simple that expert operation and care are unnecessary. Winton construction is so arranged that any working part can be investigated just by turning a hand-wheel and lifting a cover plate.

Consequently, Winton owners can make an immediate investigation of working parts, anywhere any time, without even soiling their cuffs. And on this account Winton cars seldom get into the repair shop because of postponed investigation. It is so easy to "do it now" on a Winton, that the use of a Winton, on the strength of that feature alone, means economy.

30 H. P. Type X-I-V, $2500. 40 H. P. Model M, $3500. Limousine, landaulet and runabout bodies supplied on short notice.

Photographed for Brooklyn Life by F. A. Walter.

MR. J. D. ROURKE DRIVING HIS CADILLAC CAR.
In the recent endurance run of the Long Island Automobile Club this car tied with Mr. I. C. Kirkham's Maxwell in Class A for the Brooklyn Life Cup.

Buick

The Best Low-Priced Car Built

Type C.	Type D.
2 cyl. Touring Car	4 cyl. Touring Car
Guaranteed 22 H. P.	30 H. P.
Full Set	**Full Set**
Gas and Oil Lamps	**Gas and Oil Lamps**
$1,250	$1,925

The I. S. REMSON M'f'g Co.

740-750 GRAND STREET

Sole Selling Agents for Brooklyn and Long Island

Our livery breeches for coachmen and grooms are made by the same skilled hands that make the many hundreds of gentlemen's riding breeches we sell every year.

No wonder our livery breeches fit.

Illustrated livery price list on request.

Chauffeurs' suits a specialty.

ROGERS, PEET & CO

258—842—1260 Broadway

(3 Stores)

NEW YORK

Matheson

Two Matheson Cars add perfect scores in the 4 day Sealed Bonnet Contest to the long list of Matheson performances.

Mr. R. G. Kelsey, an Amateur, drove his 1906 Matheson throughout this gruelling contest with no more need for an adjustment than on the famous trips to Chicago and Boston which he made with this same car and on which he made the best time of the year under the frightful road conditions. This in spite of the fact that the rules obliged Mr. Kelsey to make 25 miles a day more than any other car except one of a famous European make.

This performance means but little in itself, but coming as it does—so close on the other Matheson victories—the new record for gasolene cars made in the hill-climb up "Giant's Despair" and the perfect scores in other endurance runs on Decoration Day—it tells a story eloquent of Matheson reliability in any event—under all conditions.

No car in the world—not the greatest of the foreign cars—can equal the Matheson in service over American roads. The Matheson will carry seven people seventy miles an hour. You cannot purchase a more beautiful and luxurious car.

REASONABLE DELIVERIES

TOURING CARS
RUNABOUTS
35 H. P., $4,500
50 H. P., $5,000

GUARANTEED
FOR ONE
YEAR

Licensed Under Selden Patent

The Matheson Company
Of New York

Tel. 4575 Col.   1619-21-23 Broadway

Fabric and Finish.

It's the quality of that pair which has placed our livery so high.

Pricelist on request illustrating correct liveries for every man servant.

Shower-proofed dusters and other Summer motoring wear.

ROGERS, PEET & CO

258—842—1260 Broadway

(3 Stores)

NEW YORK

INTERESTED IN AUTOMOBILES?

If you are interested in motoring you should subscribe to THE MOTOR CAR. It is published monthly in the interests of automobile people and every copy is worth the yearly subscription price alone.

THE MOTOR CAR is bright and newsy and brim full of good pointers for automobile owners. For sale by all newsdealers. Single copies 10 cents.

The W. G. Pierson Publishing Co.

108 Fulton Street   New York

Photograph by F. Ed. Spooner.

CHEVROLET IN THE FAMOUS 90 H. P. "FIAT"
Just before breaking the world's track record at Morris Park.

Buick

The Best Low-Priced Car Built

Type C.
2 cyl. Touring Car
Guaranteed 22 H. P.
**Full Set
Gas and Oil Lamps**
$1,250

Type D.
4 cyl. Touring Car
30 H. P.
**Full Set
Gas and Oil Lamps**
$1,925

The I. S. REMSON M'f'g Co.
740-750 GRAND STREET
Sole Selling Agents for Brooklyn and Long Island

You Don't Take Chances

in buying your automobile from us. We will not sell unreliable cars

Pope-Toledo, 30 H. P.	$3,500
Pope-Hartford, 16 H. P.	1,600
Pope-Tribune, 12 H. P.	900
Pope-Waverley, Electric	900

A. G. Southworth

342 Flatbush Ave.,   10 Clinton St.,   811 Union St., Brooklyn

Every tire we make adds to our responsibility, because our good name is on every one. With thousands of users of

Kelly-Springfield Tires

keeping daily check on our honesty and holding us to the promises which our name implies, do you run any possible risk of being disappointed in the Kelly-Springfields you buy?

"Rubber Tired" is a little book that all drivers will like. Sent free.

Consolidated Rubber Tire Co.
39 Pine Street, New York
Akron, Ohio

RICHFIELD SPRINGS, N. Y

For your summer outing. Situated in the Otsego Hills, where the altitude is high, the air cool and bracing, the outdoor life delightful; for those suffering from gout, rheumatism and nervous diseases the sulphur baths and springs are the most efficacious in the world.

A beautifully illustrated book of 128 pages describing these and other resorts along the Lackawanna Railroad, and containing a fascinating love story, entitled "A Paper Proposal," sent for 10 cents in stamps. The edition is limited. Write for a copy to-day to T. W. Lee, General Passenger Agent, Lackawanna Railroad, New York City.

Bigger and Better Than Ever!

BROOKLYN AUTO SHOW

23rd Regiment Armory
Bedford and Atlantic Avenues

JANUARY 18th to 25th
Afternoons and Evenings (Sunday Excepted)

You who think of purchasing a car—or who plan to make your present car up-to-date with the latest accessories—will want to visit this beautiful exposition. Thirty-five leading makes in the most modern types and body designs — accessories — new body and coach work—200 cars and chassis. Special features—musical program daily.

Auspices of
Brooklyn Motor Vehicle Dealers Assn.

MILE-A-MINUTE CAR

24 Horse Power - $3,500
14 Horse Power - $2,000

Noiseless and the realization of perfection in motor vehicles. Deliveries guaranteed.

WAVERLEY ELECTRICS

Our maintenance proposition will interest you. Pre-eminently the car for city use.

Exclusive Agent

A. G. SOUTHWORTH
342 Flatbush Ave.   10 Clinton St.
BROOKLYN

Country liveries.
Whipcords for coachman, groom and chauffeur.
Correct in detail and substantially made.

Fully illustrated catalogue on request.

ROGERS PEET & CO.

258-842-1260 Broadway,

(3 Stores)

NEW YORK

In the endurance contest of the Long Island Automobile Club on May 30 and 31 a 50 h. p.

POPE-TOLEDO

stock touring car, listing at $4,250, and a 25-30 h. p.

POPE-HARTFORD

touring car, listing at $2,750, carried off all the honors. The **POPE-TOLEDO** made the best elapsed time and the **POPE-HARTFORD** the second best. Both cars made perfect scores.

The following comparative table shows what a decisive beating the **POPE-TOLEDO** and the **POPE-HARTFORD** administered to every other car in the contest:

Car	H. P.	Time First Day	Time Second Day	Total Elapsed Time
POPE-TOLEDO	**50**	**4h. 3m.**	**5h. 16m.**	**9h. 19m.**
POPE-HARTFORD	**25-30**	**4h. 51m.**	**4h. 42m.**	**9h. 33m.**
Oldsmobile	35-40	5h. 3m.	4h. 57m.	10h. 0m.
De Luxe	50-60	6h. 21m.	4h. 52m.	11h. 13m.
Winton	40	5h. 48m.	6h. 2m.	11h. 50m.
Pierce	40-45	6h. 18m.	6h. 1m.	12h. 19m.
Columbia	24-28	5h. 37m.	6h. 46m.	12h. 23m.
Packard	30	6h. 5m.	6h. 18m.	12h. 23m.
Welch	50	5h. 41m.	6h. 48m.	12h. 29m.
Franklin	20	6h. 15m.	6h. 18m.	12h. 33m.
Matheson	40-45	6h. 40m.	6h. 32m.	13h. 12m.
Aerocar	40	6h. 29m.	7h. 2m.	13h. 31m.
Dolson	60	6h. 43m.	7h. 6m.	13h. 49m.
Queen	30	7h. 14m.	8h. 0m.	15h. 14m.
Thomas	40	5h. 40m.	9h. 40m.	15h. 20m.
Haynes	30	6h. 51m.	Did not finish.	

The first day's run was 139 miles, and the second day's run was 154.6 miles.

The **POPE-HARTFORD**, carrying four passengers, was first at every control and first at the finish. Its elapsed time for the run home was the best made, while on the out run it was beaten only by the **POPE-TOLEDO**.

The **POPE-HARTFORD'S** total elapsed time was 27 minutes better than that of the Oldsmobile, which finished second in the class.

The **POPE-TOLEDO'S** total elapsed time was 1 hour 54 minutes better than that of the De Luxe, its nearest rival (barring the **POPE-HARTFORD**).

Not an adjustment of any kind was made to the engine of either car.

A. G. Southworth & Co., Inc.

1733 Broadway, N. Y.   342 Flatbush Ave., B'klyn

Metropolitan Agents for all the Pope Cars

TOURING TIME is Rambler TIME

THE SUCCESS of your trip is entirely dependent upon the reliability of your car. Then, as at no other time, is a capacity for steady service under all conditions of such paramount importance.

The production of a car of absolute dependability has ever been the primal object of the Rambler factory, and the thousands of these cars now in constant service are ample proof of successful efforts.

Built in seven models, $1,200 to $3,000

Main Office and Factory, - - Kenosha, Wis.

Branches:

Chicago, 302-304 Wabash Ave.
Boston, 145 Columbus Ave.
San Francisco, 31 Sanchez St.
New York Agency, 38-40 W. 62nd St.
Milwaukee, 457-459 Broadway
Philadelphia, 242 N. Broad St.
Representatives in all leading cities.

Thomas B. Jeffery & Company

Model 15, $2,500

The Kelly-Springfield Tire

There are facts about rubber tires that are important to every user of carriages. The value of a good tire such as the **Kelly-Springfield Tire** is a real and tangible value. It adds to the comfort and service of your carriage. It adds to its style and appearance. It is used and endorsed by all good judges. The better grade carriages a manufacturer makes the more likely he is to put Kelly-Springfield Tires on them.

A booklet entitled "Rubber Tired" is sent to every one for the asking.

Consolidated Rubber Tire Co.
39 Pine Street, New York
Akron, Ohio

The Great STRENGTH, DURABILITY and RESILIENCE possessed by the

Healy Leather Tire

are occasioned by reason that *leather*, *rubber* and *fabric* each are placed exactly where they are needed to meet the various strains and wear, under the most severe conditions.

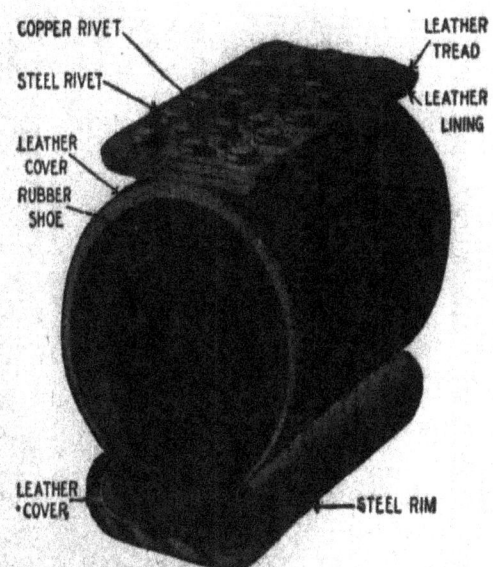

SECTION OF CLINCHER TIRE LEATHERIZED

No Punctures　　　　　　　　　　No Blow-outs
No Rim-cutting　　　　　　　　　 No Skidding

WE GUARANTEE OUR TIRES
EXPERT REPAIRING OF EVERY DESCRIPTION

We have much more to say.　　　　　*Let us tell it to you.*

Healy Leather Tire Co.

HEALY BUILDING
88-90 Gold St., New York City.

Baker Electrics

A complete line of

| BEVEL GEAR | SHAFT DRIVE |

Models now on exhibit at our show room.

The adoption of this new type of bevel drive is the greatest single advance ever made in Electric Motor Car construction.

Our 1910 catalogue will be mailed on request.

Demonstration can be arranged at your convenience.

Baker Vehicle Co.

Phone 2830 Columbus
1788 Broadway, - New York

Photographed for Brooklyn Life by F. A. Walter.
MR. C. P. SKINNER'S 20 H.P. MITCHELL RUNABOUT ON RIVERSIDE DRIVE.

"IF YOU CAN'T COME TO ME I'LL COME TO YOU"

The "Maxwell"

Model "H. C." 20 H.P. Fully Equipped

$1,450

The Car which has made the name of its designer famous the world over.

"AN IDEAL FAMILY CAR"

I. C. Kirkham

Exclusive Distributor for Long Island

1060 Bedford Avenue, Cor. Clinton Place, Brooklyn
Telephone 4300 Bedford

All Model 10 Cars are equipped with magneto, tools and five lamps.

All Model 10 Cars have 4-cylinder motors rated at 22½ horse power A.L.A.M.

THE MODEL 10 Buicks are designed and built to run at a lower cost of upkeep, mile for mile, than any other cars, regardless of price, horsepower or number of cylinders. They have won more hill climbs, speed and endurance contests than any other low-priced cars.

THE MODEL 10 Buick is not an experiment, not a new first year car that will develop weaknesses. Time and experience have perfected it. In the hands of 17,000 private owners who have run their cars from 1,000 to 60,000 miles they have proven, one by one, for all-around use the best built and most popular light car in America.

BUICK cars are the best automobile investment because they bring a better percentage of the original price when sold second-hand than any other make of car, barring none. *Ask your neighbor—he has one.*

Policy

The highest aim of the entire sales organization of the Buick Motor Company is to take proper care of its customers. With this thought in mind we solicit your valued trade.

BUICK MOTOR COMPANY

NEW YORK
Broadway at 55th St.

BROOKLYN
42 Flatbush Avenue

NEWARK
222 Halsey St.

You See Them Everywhere

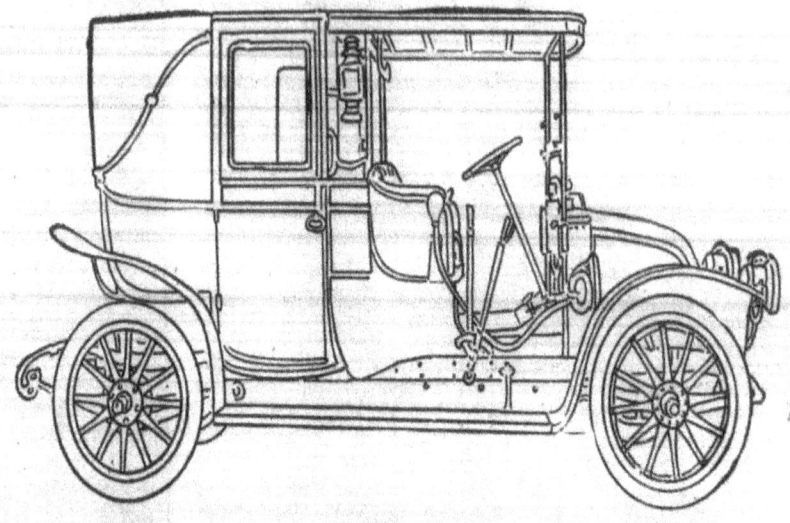

Renault

The New 1910 Town Car

Direct from Paris _____ The Latest Fashion

| COLLAPSIBLE LANDAULET | . . . | $4,250 |
| EXTENSION LANDAULET IMMEDIATE DELIVERY | . | 5,000 |

For immediate delivery, a brougham by Kellner, of Paris, a duplicate of the car built for King Edward of England **$5,250**

THE CAR OF SILENCE THE CLASSIEST OF CARS

CHASSIS SPECIFICATIONS:—Curved frame of pressed steel, 12-16 h. p. Renault 4-cylinder motor, mechanically operated valves, Bosch magneto, automatic carburetor, 3 speeds forward, direct drive on high gear, reverse, shaft drive, thermo syphon cooling system.

NO COIL—NO BATTERY—NO WATER PUMP.

RENAULT FRERES SELLING BRANCH

PAUL LACROIX, Gen. Mgr., 1,776 Broadway (Corner 57th Street), New York

Four Striking Types of Racing Cars That Have Attracted Wide Attention.

D. G. BUCK OF CHICAGO DRIVING THE FAMOUS APPERSON "JACKRABBIT."

THE MAJA, ENTERED FOR BRIARCLIFF TROPHY BY MR. JOHN BROWN

THE AMERICAN ROADSTER ENTERED IN THIS WEEK'S SAVANNAH RACE.
The driver is Fred I. Tone, the designer of the car.

MR. HARRY LEVEY DRIVING HIS 120 H.P. HOTCHKISS RACER.
This car was entered by Mr. Levey in the Florida races of this month.

Photographed for Brooklyn Life by F. A. Walter.
MRS. DORA CARRINGTON'S 35-45 H.P. RENAULT, WITH BREWSTER LIMOUSINE.
Messrs. Paul Lacroix and H. V. Kibbe of Renault Frères are testing this car before delivery.

through the Brooklyn Armstrong Association two of the   ACCORDING to all accounts the musicale which Mrs. Her-

Photographed for Brooklyn Life by F. A. Walter.

MR. J. B. CLEWS'S 50 H.P. MARTINI LIMOUSINE, MOORE & MUNGER BODY.

MESSRS. SIDNEY BOWMAN AND CLEMENT IN THE VICTORIOUS CLEMENT-BAYARD RACER.

ELLENBECK & MULLER
AUTOMOBILE GARAGE

912 BEDFORD AVE.   TELEPHONE 4647 WMBG.
BROOKLYN

We invite you to inspect our Garage in reference to Storage, Repairs, etc. Our repair shop is under our own supervision. We also paint automobiles. Our price is right and our work perfect.

AGENTS FOR THE

SIX MODELS
35 h. p. $2,750   to   96 h. p. $7,500

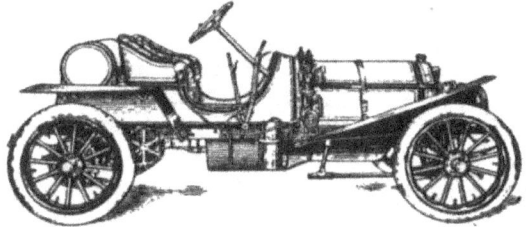

THE APPERSON JACKRABBIT
THE FASTEST CAR FOR ITS POWER IN THE WORLD

Long and thorough demonstrations upon request. We carry a good line of slightly used cars.

FORDS   AUTO-CAR   CHADWICK
CLEMENT-BAYARD

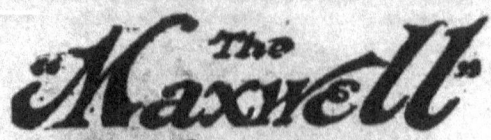

The "Maxwell"

A MODEL TO MEET EVERY DEMAND

"L.C."
Runabout
14 H.P.
$825.

"H.C."
Touring Car
20 H.P.
Fully
Equipped
Five
Passengers
$1,450

"K"
Gentleman's
Roadster
Folding
Rumble Seat
4 Cyl. 28 H.P.
$1,750.

"D"
Touring Car
4 Cyl. 28 H.P.
$1,750.

I. C. KIRKHAM, Exclusive Distributer for Long Island
1060 BEDFORD AVE., Cor. Clinton Place, BROOKLYN
Telephone 4300 Bedford. OPEN EACH EVENING

STEVENS-DURYEA SIX CYLINDER CARS

¶While the attention of the Stevens-Duryea Company, in the main, will be concentrated on Sixes, they are bringing out a new four-cylinder car, which they consider about the limit of horsepower, size and speed that should be embodied in a Four. They are building this car to supply a demand for a Four of medium power and of moderate price. Many other makers of Fours are building a two-cylinder to supply the demand for that type.

¶They firmly believe that the six-cylinder car represents the acme of motor car enjoyment and any person paying more than $2750 for a car should get a Six.

I. M. ALLEN CO.
116 So. Portland Ave.
Brooklyn, N. Y.

Tel. 4026 Prospect

Manufactured by Stevens-Duryea Company, Chicopee Falls, Mass., U. S. A.

BAKER

ELECTRIC VEHICLES

BAKER ROADSTER

BAKER QUEEN VICTORIA

BAKER MOTOR VEHICLE CO.
OF NEW YORK
1790 BROADWAY

STUDEBAKER
(the Great Independent)

Announces a new

President
Straight Eight

100 horsepower

$1985
f. o. b. factory

See this luxurious new Studebaker at the Show!

FORD
"S" ROADSTER,
$750.00 F. O. B. Detroit
WE ARE NOT TALKING PRICE, WE ARE SHOUTING QUALITY

A DEMONSTRATION WILL CONVINCE YOU

MODELS COMING
A light Touring Car, 22-24 H. P., equipped with magneto; price about . . . . . **$850.**
A Landaulet same specifications; price about . **$1,000.** 6 Cylinder Touring Car . **$2,800.**

DISTRIBUTERS:
BISHOP, McCORMICK & BISHOP, Inc.,
Halsey Street, Just off Bedford Avenue, Brooklyn
Telephone 3902, 3903 Bedford

TOLEDO XVII	HARTFORD "30"
$4750.00	$2750.00

THE 4 POPES

That cover the Automobile proposition completely from the luxurious seven passenger touring car to the inexpensive runabout. We invite investigation and back up every statement with an absolute guarantee. The **largest** supply department. The **best** equipped **shops** and **garages** in the city.

JOHN W. SUTTON

342 Flatbush Avenue

10 Clinton Street - 811 Union Street

BROOKLYN

A. W. Blanchard, Manager

TRIBUNE	WAVERLEY
$1500.00	$1425.00 up. Electric

This is the "SHOW ME"

$1250 *Mitchell*

4 Cyl., 20 H. P. Rumble Seat Roadster

A "BEAUTY" in style, finish and CONSTRUCTION--- as for PERFORMANCE on the road---let us "show you" and PROVE to you how the Mitchell "makes good."

Let us take you out for a 20 or 40 mile ride, so you can see just what the Mitchell can do, and CONVINCE YOURSELF of its great value

Call us up (4647 Williamsburg) and say "Show Me" the Mitchell and we will be ready any time you say. No obligation. If you are interested we'll be glad to.

Mitchell Motor Co., of N. Y.
912 Bedford Avenue

MITCHELL MOTOR CAR CO., Racine, Wis., Member Am. Motor Car Mfrs'. Assn.

Ormond-Daytona Beach Races
March 3—1908

160 Mile Stock Chassis Race
  1st. Benz . . . . . . 80 Horse Power
  2nd. CLEVELAND (Pathfinder) 40 Horse Power
  Other Entrants . . . . . Packard, Haynes, Thomas, Christie and Allen-Kingston.

NOTE:

The Benz was protested owing to being 350 pounds over-weight and excessive cylinder displacement. The CLEVELAND car that won second place was the same car that a week before finished the Jacksonville-Miami—371 Miles endurance Run through trackless Florida swamps, sand barrens and turpentine woods in 5 days carrying 1,000 pounds of baggage and five passengers. This without a break down or a replacement of any kind beyond a broken front spring and you can buy the same kind of a car for

$3,500

Touring Car or Runabout

The Cleveland Motor Car Company
1659 BROADWAY,   NEW YORK CITY

Philadelphia Branch   Chicago Branch
236 North Broad Street   1218 Michigan Avenue

RENAULT
"The Car"

HOLDS WORLD'S RECORDS
100 miles in 1 hour 12 minutes 56 1-5s
24 hour   1079 miles

Speed and Endurance are the best proofs of

RELIABILITY

RENAULT PRICES:

50-60 H.P. 6-CYLINDER CHASSIS	$8,250
35-45 H.P. 4-CYLINDER RUNABOUT	7,250
35-45 H.P. 4-CYLINDER TOURING	7,750
35-45 H.P. 4-CYLINDER LIMOUSINE OR LANDAULET	8,250
20-30 H.P. 4-CYLINDER RUNABOUT	5,750
20-30 H.P. 4-CYLINDER TOURING	6,250
20-30 H.P. 4-CYLINDER LIMOUSINE OR LANDAULET	6,750
14-20 H.P. 4-CYLINDER CHASSIS	4,250
10-14 H.P. 4-CYLINDER TAXIMETER CAB	3,950
8-10 H.P. 2-CYLINDER TAXIMETER CAB	2,750

RENAULT FRÈRES SELLING BRANCH

Broadway and 57th Street,   -   New York City

PAUL LACROIX, General Manager.   'Phone 3004 Columbus
CHICAGO BRANCH, 1549 Michigan Avenue
SAN FRANCISCO BRANCH: 316-322 Van Ness Avenue
PHILADELPHIA AGENCY: Prescott Adamson, Broad and
  Spring Garden Streets
BOSTON AGENCY: A. Cutler Morse & Co., Motor Mart,
  Park Square

MR. J. W. MEARS DRIVING HIS 1908 ACME SEXTUPLET ON THE CONEY ISLAND BOULEVARD.

THE
ACME
SEXTUPLET

Price, [$4,500, Complete

is the **ACME** of **RELIABILITY, SPEED, STYLE,** and **COMFORT,** with cost of maintenance extremely low. Write or Telephone **1454** Flatbush for demonstration at any time.

J. W. MEARS
7-9 Ocean Parkway.

GARAGE WITH BEST EQUIPMENT AND SKILLED MECHANICS

Locomobile

TYPE E 1908

The Car of Durability that has Stood the Test of Time

Every necessary feature requisite to produce the Perfect Touring Car is found in the

Type "E" and Type "I" Models for 1908
DEMONSTRATIONS

The I. S. REMSON M'f'g Co.
Sole Agents for Brooklyn and Long Island
Garage—754-760 Bedford Avenue

MERCANTILE MOTOR CAR CO. (Inc.)

Successors to Long Island Motor Car Co.   Telephone, 2538 Prospect

General Machinists   **Automobiles Stored**

Expert Repairing of all Cars   Repaired and sold on commission

FIRST-CLASS CARS Fully Equipped for Renting   Dealers in Second-hand Automobiles

368 Cumberland Street,   **Brooklyn, N. Y.**

P E N N S Y L V A N I A

THERE ARE NO HILLS

The way to "size up"

The way to "size up" a proposition is to be "on the ground." If you will let us know where your "ground" is we will get a "Pennsylvania" on it—i. e., if you think of buying a really good car—advertising rhetoric don't make "the goods"—there is nothing so bad but what good *can be said* of it.

50 H. P.
114 W. B.

$2,800

Grant Square Auto Co.
Brooklyn and L. I. Distributers
1378 Bedford Avenue

Photographed for Brooklyn Life by F. A. Walter.
MR. BURTON T. BISHOP DRIVING HIS SIX-CYLINDER FORD TOURING CAR.
The other occupant of the car is Mr. John McCormick.

Mr. Charles A. Carlson driving his Winton Six-teen-Six Runabout. Mr. Chas. D. Smith in front seat and Mr. Chas. B. Shanks in rumble.

QUINBY
ALUMINUM BODIES

LIMOUSINE,
LANDAULET,
TOURING,
RUNABOUT.

COMPLETELY EQUIPPED
MOTOR REPAIR
DEPARTMENT.

PANHARD AND SIMPLEX CARS

Early Delivery

J. M. QUINBY & CO.

NEWARK, N. J.
Adjacent Lackawanna Station.

Matheson

Designed to compete only with the finest and highest-priced cars of foreign or domestic make.

Noted for its never-failing reliability, its strength, power and lasting durability.

Luxurious in every appointment.

MATHESON owners are enthusiatic MATHE-SON advocates.

50 H. P.—$5,500. IMMEDIATE DELIVERIES

A. G. SOUTHWORTH CO., Inc.
1733 BROADWAY, NEW YORK

Distributory for parts and the best facilities in the East for overhauling and refinishing all models of Pope-Hartford, Pope-Toledo and Matheson cars.

Healy Leather Tires
Healy Rapid Removable Rims
Healy Leather Tire Co.

90 Gold Street   New York   1906 Broadway

WATCH YOUR AUTOMOBILE

And see that your CHAUFFEUR uses only

BELDEN'S AUTO-VALVONE OILS

The Highest Grade Lubricant Known for Hydro-Carbon Engines
No Corrosion Possible

A. G. BELDEN & CO.
Sole Manufacturers
145 Maiden Lane, New York

MR. CHARLES F. BATT DRIVING HIS PENNSYLVANIA CAR.

ONE OF THE NEW 75 H.P. 6-CYLINDER GEARLESS CARS, MODEL 75.

The New Stevens-Duryea Model X

Because of a demand for a larger and roomier car than the present twenty-horse-power, four-cylinder Model R, at practically the same price, the Stevens-Duryea Company have produced this new Model X. It embodies many special improvements and refinements. The cylinders are cast in pairs and the crank-shaft turns in Babbitt bushings. The valves are on one side of the engine, thus necessitating but one cam-shaft. The spur gears driving the pump shaft and the cam-shaft are operated from the clutch end of the power plant and being entirely enclosed are very quiet.

The oiler is enclosed under the hood and is direct connected. The control is by spark and throttle levers, located on a sector on the steering wheel, but not revolving with it. There is also a foot acceleration for the throttle.

The body is roomy and comfortably seats five. The cross-spring at the rear end of the chassis gives a smooth, easy motion when on the road. An actual road speed of from forty-five to fifty miles an hour can be attained.

The frame is of chrome nickel steel; wheel base one hundred and twenty-four inches; body of rolled sheet aluminum and mud guards also of aluminum. Complete, the car weighs twenty-six hundred pounds and is furnished with acetylene headlights, generator, two oil side lamps, one oil tail lamp, coat rail, foot rest and tool case. Price, $2,750. Further information or particulars of this new car will gladly be given by the I. M. Allen Company of 116 South Portland Avenue, the Brooklyn and Long Island distributors of Stevens-Duryea automobiles.

Photographed for Brooklyn Life by F. A. Walter.
MR. JOHN W. SUTTON DRIVING HIS POPE-TOLEDO CAR IN PROSPECT PARK.

Photograph by Spooner & Wells.
M. G. BERNIN AND THE RENAULT THAT WON THE HUNDRED MILE RECORD.

Photograph by Spooner & Wells.

A MIDWINTER RIDE AMONG THE PINES IN A GARFORD CAR.

New 1908 Models of Some Well Known American Automobiles.

A CHADWICK TOURING RUNABOUT IN FAIRMOUNT PARK, PHILADELPHIA.

Photograph by N. Lazarnick.
MR. J. C. KIRKHAM DRIVING HIS 28 H.P. MAXWELL CAR, L.I.A.C. RUN.

Photographed for Brooklyn Life by F. A. Walker.
MR. W. H. KOUWENHOVEN DRIVING A LOCOMOBILE IN PROSPECT PARK.
The owner of this car is Mr. James Doig, Bayville, L.I.

MR. J. P. LOGAN, JR. IN HIS MATHESON CAR IN CENTRAL PARK.
In the rear seat are Mr. A. G. Southworth and Mr. H. W. Jones.

CAPTAIN JENKINS OF THE SEVENTY-FIRST IN HIS ACME CAR AT CREEDMOOR.

MODEL AA.

12 H. P.   "*Maxwell*" $600

A Runabout with a Reputation—NOW.

The Model "AA" MAXWELL has made its mark in the Automobile world—Reliability, Efficiency and just return as an investment have been credited to it by a discriminating public.

YOU CANNOT IGNORE PROOF OF FACT

Phone
4300 Bedford

I. C. KIRKHAM

Guaranteed
Deliveries

1060-1062 BEDFORD AVENUE

MEMBER BROOKLYN MOTOR VEHICLE ASSOCIATION

White Motor Cars

Reliable

Comfortable

White Cars Are Good Cars.
None Better at Any Price.

Agents for Brooklyn & Long Island

Repairing   Upholstering
Trimming   Painting
A Specialty

BOROUGH AUTOMOBILE CO.
Show Rooms, 1285 Bedford Ave.
Garage, 679 McDonough St.

Reliability   Luxury   Refinement

Studebaker "40"

Touring, Enclosed and Roadster body

Studebaker

ELECTRICS

For the Family's Use   **VICTORIAS**
(Interior Driven)   **Coupes**
    LANDAULETS

Carpenter Motor Vehicle Co.
BROOKLYN AGENTS
1239-41-43 FULTON STREET
Telephone 300 Bedford for demonstration.   Open Evenings.

Photograph by Spooner & Wells.
THE FIAT CYCLONE THAT WAS RUN BY CEDRINO AT ORMOND.

Mr. John G. Gasteiger, of 1111 Dean Street, in his 1910 Moon "30".

AUTO SHOW

GRAND CENTRAL PALACE

OPENS TOMORROW AT 2 P.M.

See What's New at This Great Show!

JANUARY 7–14 10 A.M. TO 10.30 P.M.
(SUNDAY EXCEPTED)

CARS, ACCESSORIES, COMMERCIAL VEHICLES — Hundreds of innovations for 1933 — Never before such car values.   ADMISSION 75¢

FOUR PASSENGER TOURABOUT

30 H. P. 4 cylinders, cast in pairs $4\frac{1}{2} \times 5$ inches with magneto and 5 lamps

PRICE $1,750

Without exception the strongest, safest and easiest riding car sold at a moderate price.

BUICK MOTOR CO.

NEW YORK   **BROOKLYN**   **NEWARK**
55th St. & B'dw'y   42 Flatbush Ave.   222 Halsey St.

The new Renault 1910 Town Car

THE INCOMPARABLE
WHITE
THE CAR FOR SERVICE

A TOURING CAR IN TRUE SENSE OF THE TERM

The man who gets the most pleasure from his touring car is not the man who limits his touring to the macadam roads; for the most interesting sections of the country and those of the greatest natural beauty lie, for the most part, beyond the regions of improved highways. For that reason, there is no quality of a motor car more important than the ability to traverse bad roads.

In unique degree, the White possesses the qualities of a "bad roads" car. Owing to the perfect flexibility of the engine, the White tourist can accommodate the speed of his car, yard by yard, to the condition of the road, speeding up on each little stretch of good road, and slowing down for each hole and "thank-ye-ma'am"—without shifting of gears or any manipulation except of the throttle. The tremendous pulling power of the White engine under all conditions means immunity from getting stuck in the mud or sand. Running through deep water, as in fording streams, is easy for a White. And as for climbing grades in mountainous regions—there is no other machine which can approach the White in hill-climbing qualities.

Drive a White Steamer and see the country

THE WHITE COMPANY
CLEVELAND, OHIO.

NEW YORK CITY, Broadway at 62nd Street
SAN FRANCISCO, 1460 Market Street
PHILADELPHIA, 629-33 North Broad Street
BOSTON, 320 Newbury Street
CHICAGO, 240 Michigan Avenue
CLEVELAND, 407 Rockwell Avenue
PITTSBURG, 138-148 Beatty Street

WANTED—A RIDER AGENT IN EACH TOWN

and district to ride and exhibit a sample Latest Model **"Ranger"** bicycle furnished by us. Our agents everywhere are making money fast. *Write for full particulars and special offer at once.*

NO MONEY REQUIRED until you receive and approve of your bicycle. We ship to anyone, anywhere in the U. S. *without a cent deposit* in advance, *prepay freight*, and allow **TEN DAYS' FREE TRIAL** during which time you may ride the bicycle and put it to any test you wish. If you are then not perfectly satisfied or do not wish to keep the bicycle ship it back to us at our expense and *you will not be out one cent.*

FACTORY PRICES We furnish the highest grade bicycles it is possible to make at one small profit above actual factory cost. You save $10 to $25 middlemen's profits by buying direct of us and have the manufacturer's guarantee behind your bicycle. **DO NOT BUY** a bicycle or a pair of tires from *anyone* at *any price* until you receive our catalogues and learn our unheard of *factory prices* and *remarkable special offers* to **rider agents.**

YOU WILL BE ASTONISHED when you receive our beautiful catalogue and study our superb models at the *wonderfully low prices* we can make you this year. We sell the highest grade bicycles for less money than any other factory. We are satisfied with $1.00 profit above factory cost. **BICYCLE DEALERS,** you can sell our bicycles under your own name plate at double our prices. Orders filled the day received.

SECOND HAND BICYCLES. We do not regularly handle second hand bicycles, but usually have a number on hand taken in trade by our Chicago retail stores. These we clear out promptly at prices ranging from **$3 to $8 or $10.** Descriptive bargain lists mailed free.

COASTER-BRAKES, single wheels, imported roller chains and pedals, parts, repairs and equipment of all kinds at *half the usual retail prices.*

$8.50 HEDGETHORN PUNCTURE-PROOF SELF-HEALING TIRES $4.80
A SAMPLE PAIR TO INTRODUCE, ONLY

The regular retail price of these tires is $8.50 per pair, but to introduce we will sell you a sample pair for $4.80 (cash with order $4.55).

NO MORE TROUBLE FROM PUNCTURES
NAILS, Tacks or Glass will not let the air out. Sixty thousand pairs sold last year. Over two hundred thousand pairs now in use.

DESCRIPTION: Made in all sizes. It is lively and easy riding, very durable and lined inside with a special quality of rubber, which never becomes porous and which closes up small punctures without allowing the air to escape. We have hundreds of letters from satisfied customers stating that their tires have only been pumped up once or twice in a whole season. They weigh no more than an ordinary tire, the puncture resisting qualities being given by several layers of thin, specially prepared fabric on the tread. The regular price of these tires is $8.50 per pair, but for advertising purposes we are making a special factory price to the rider of only $4.80 per pair. All orders shipped same day letter is received. We ship C. O. D. on approval. You do not pay a cent until you have examined and found them strictly as represented.

We will allow a **cash discount** of 5 per cent (thereby making the price **$4.55 per pair**) if you send **FULL CASH WITH ORDER** and enclose this advertisement. You run no risk in sending us an order as the tires may be returned at OUR expense if for any reason they are not satisfactory on examination. We are perfectly reliable and money sent to us is as safe as in a bank. If you order a pair of these tires, you will find that they will ride easier, run faster, wear better, last longer and look finer than any tire you have ever used or seen at any price. We know that you will be so well pleased that when you want a bicycle you will give us your order. We want you to send us a trial order at once, hence this remarkable tire offer.

Notice the thick rubber tread "A" and puncture strips "B" and "D," also rim strip "H" to prevent rim cutting. This tire will outlast any other make—SOFT, ELASTIC and EASY RIDING.

IF YOU NEED TIRES don't buy any kind at any price until you send for a pair of Hedgethorn Puncture-Proof tires on approval and trial at the special introductory price quoted above; or write for our big Tire and Sundry Catalogue which describes and quotes all makes and kinds of tires at about half the usual prices.

DO NOT WAIT but write us a postal today. **DO NOT THINK OF BUYING** a bicycle or a pair of tires from anyone until you know the new and wonderful offers we are making. It only costs a postal to learn everything. Write it **NOW.**

J. L. MEAD CYCLE COMPANY,   CHICAGO, ILL

Buick Model 10 Surrey, $1,050

This is a Buick Advertisement full of facts, information and honest comparisons. There is sincerity in every word.
We spend very little money advertising Buick Cars in papers, magazines or catalogues, but last year's registrations prove that we sold more cars than any one company. The reason is in the car itself. Our customers are our best salesmen.

Buick Advantages

THE DESIGNING OF THE PRESENT BUICK CARS WAS A SEVEN-YEAR TASK. We had the experience and suggestions of 32,000 Buick owners to help us, and any good suggestion *has been* and *is* welcome and recognized, no matter whether it comes from a Buick owner, the General Manager of the Company, a Branch Manager, Agent, designer, tester, chauffeur or office boy.

WE HAVE THE MOST PRODUCTIVE FACTORY. Our latest schedule calls for 40,100 Buick Cars for 1910. The fact that we have a market for this enormous output proves that the Buick has made good. We have the most representative branch houses and the greatest sales organization of any company that deals in automobiles. When our New York salesrooms are complete, they will be the largest in America. We have the largest warehouse and stock room in New York City, where you can secure or inspect every part that goes into a Buick Car. It never pays to purchase a car of a small manufacturer who may go out of business at any time and leave you without parts —a car that has no second-hand value. You can get a Buick part and find a mechanic who knows the car in every hamlet. This is one reason why Buick Cars are a better investment and bring better prices when sold second hand than any other make of car, BARRING NONE.

WE USE FULL ELLIPTIC SPRINGS AND LARGE TIRES, therefore the car rides easily. We give magneto and gas lamps without extra charge. We furnish motors that show *actual power*—not catalogued horse power. We guarantee that the cars we deliver will take you up Miller Avenue hill easily on high gear. We are the only manufacturers selling cars for less than $3,000 that catalogue their cars less than the A. L. A. M. rating.

WE KNOW THAT OUR FRAMES ARE STRONG because we have delivered over 32,000 cars, and have never had trouble. We know that our steering parts are safe, as we use a larger worm and sector than any medium priced car. We know our wheels and axles are the best made, because we are the largest high grade automobile wheel and axle manufacturers in the world. We know that our brakes are ample in size and easily applied, because they will lock our wheels at any time under any conditions, which is all any brakes can do. We know that our 1910 transmission is the strongest in any car sold for less than $4,000. We know that our springs are right, because they have never given trouble.

All this means that Buick Cars have ample power, endurance and tire equipment for everyday service, enjoyable touring in the country or vacation trips in the mountains.

Buick Records

All the World Loves a Winner

BUICK RECORDS ARE THE GREATEST EVER MADE BY A STOCK CAR REGARDLESS OF PRICE, HORSE POWER OR NUMBER OF CYLINDERS. In 1909 Buick Cars won 204 firsts; in two years they have won 316 firsts. No other make of car during the history of motordom has equaled this record. A Buick holds the American Road Race Record and World's Stock Car Record of 70 miles an hour, made at Riverhead, L.I. A Buick holds the 100 mile record for one mile track, time 101 minutes 25 seconds, made at Dallas, Tex. A Buick holds the 200 mile American Track Record, average speed 72 miles an hour, made at Atlanta, Ga. A Buick holds the 250 mile American Track Record, made at Indianapolis, Ind. A Buick holds the world's 12-hour mile track record of 624 miles, made at Brighton Beach. A Buick won the two longest and most severe road races ever held in America— the 396 mile Cobe Trophy Race and the 482 mile Los Angeles to Phoenix Race. A Buick Car driven by Mr. Wm. Oldknow, its owner, in the New York Herald-Atlanta Good Roads Run secured a perfect score. On arriving at Atlanta, Mr. Oldknow won the 10 mile amateur free-for-all race in 8 minutes, 52 seconds.

A BUICK CAR won its class in New York Auto Trade Association one gallon efficiency test, covering 28.2 miles. Buick Cars won seven of the ten National Stock Chassis races at Lowell, leaving three other events for 20 manufacturers and 30 drivers of national reputation.

THE BUICK SPECIALTY was winning long distance races, where great endurance was required. Buick Cars won more road races, track races and hill climbs than any other car; they won a greater percentage of the contests entered (91 per cent.) than any other car, won more important events than any other car and at a greater average speed in miles per hour.

BUICK CARS ARE ACKNOWLEDGED ALL-AROUND CHAMPION STOCK CARS. Other companies have claimed the stock car championship that do not hold a world's record on road or track, that never won a big race when high-priced, high-power cars have entered. A comparison of records on file at our office will prove these statements.

Buick Policy — The highest aim of the entire sales organization of the Buick Motor Company is to take proper care of its customers. With this thought in mind, we solicit your valued trade.

Remember—We demonstrate on Miller Avenue Hill on High Gear. Ask Others to Do Likewise

BUICK MOTOR COMPANY

NEW YORK
Broadway & 55th St.

BROOKLYN
42 Flatbush Avenue

NEWARK
222 Halsey Street

FLANDRAU & CO.
AUTOMOBILES

Factory and Repository

406, 408, 410, 412 Broome Street
NEW YORK

SELLING AGENTS
For the renowned French car

BRASIER

Formerly Richard Brasier
(Licensed Under Selden Patent)

THE WINNER IN

Gordon-Bennett Cup Races
Hill Climbing Contests
Speed and non-stop tests
and Marine Contests

Better than ever and built for American Roads

All sizes for Town and Touring

We offer chassis only, or fitted with our unexcelled

COACH WORK

Repairs to Bodies or Motors

Also agents for the
BEST ELECTRIC VEHICLE—
The RAUCH & LANG

SILENCE
COMFORT

TOWN CAR

20 H.-P. Town Car

THE Peerless Town Car, in limousine and landaulet models, is designed especially to meet the requirements of city and suburban business and social travel. The lighter construction, modified power, shorter wheel base, and left-hand drive, are conveniences particularly adapted to many usages not demanding a large car. Wherever possible, the grace and refinement of these models has been enhanced.

This new car may now be seen at our Salesrooms. We are prepared to accept a limited number of orders for immediate delivery.

THE PEERLESS MOTOR CAR CO. OF N. Y.
1760 Broadway, at 57th Street
PEERLESS GARAGE & SALES CO.
1525 Bedford Ave., near Eastern Parkway

Licensed under Selden Patent

THE PALMER & SINGER closed cars embody all essentials of successful town cars—silence, comfort, easy riding and easy handling, beside exceptional beauty of finish and grace of line.

LIMOUSINE and landaulet bodies fitted on our Four-Thirty chassis have proven especially satisfactory to our clientele. The engine is so flexible that it is peculiarly adapted to town service. The landaulet makes a happy combination of open and closed car and has won signal favor.

The Four-Thirty motor is silent and smooth running, and has developed plenty of power in road tests. The low drop chassis frame, giving easy ingress and egress, makes it particularly satisfactory for shopping, calling or similar uses. The bodies with which it is equipped are smart and commodious, and include all of the nice accessories which distinguish the private equipage.

We especially invite you to compare our specifications with those of other well-known cars. You cannot buy similarly high quality for less than $1,500 above our price. Isn't this worth investigation?

We will demonstrate our cars to you at any time, sending to your house or hotel by appointment. Address Sales Manager B.

IMMEDIATE DELIVERIES

Palmer & Singer
Manufacturing Co.

Salesroom, 1620 Broadway, (at 50th Street)

FACTORY—LONG ISLAND CITY

BROOKLYN—ALLEN SWAN CO.
1384 BEDFORD AVE.

Licensed Under the Selden Patent

ALL OUR CARS GUARANTEED FOR ONE YEAR

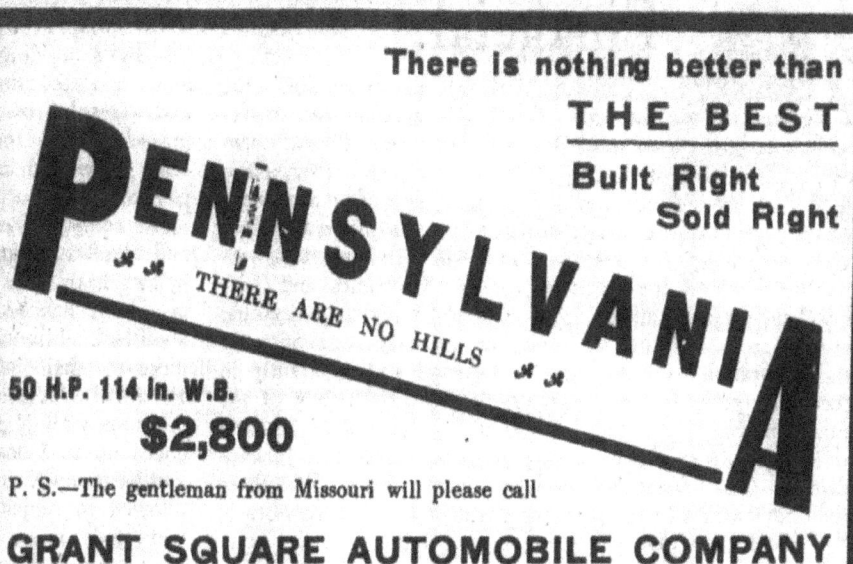

Photograph by N. Lazarnick.
M. G. BERNIN DRIVING THE 60 H.P. RENAULT RACER, NOW AT ORMOND.

WHY PAY MORE WHEN LESS WILL PURCHASE EQUALLY AS MUCH?

The "**Maxwell**"

"LIGHT FOUR"

Speed
    Endurance
        Power

$1,750.

Model D. 24-28 Horsepower Touring Car
Model K " " RUNABOUT

The same Type Engine which made the Confetti Car of the last Glidden Tour a wonder to all who could follow the trail it blazed

IF YOU CAN'T COME TO ME I'LL COME TO YOU

I. C. KIRKHAM
Exclusive Long Island Distributor

Telephone { 4300 / 4301 } BEDFORD

1060 Bedford Avenue

THE
ACME
CARS

ARE THE MOST **RELIABLE** IN THE WORLD FOR ALL **TOURING** CONDITIONS. SPEED, STYLE, COMFORT, UNSURPASSED.

J. W. MEARS

7-9 Ocean Parkway,   'Phone 1454 Flat.

Personal demonstration at any time by telephoning

WALTER HALE DRIVING HIS STUDEBAKER RUNABOUT ON THE ROAD TO TUXEDO.

MR. CHARLES A. SKINNER, JR., IN HIS NEW SKIMABOUT.
This 28-35 h.p. Palmer-Skinner car is the first of its type to be seen in New York.

FIRST PLACE to
Diamond

BEST CASINGS — **TIRES** — **BEST TUBES**

In BRIARCLIFF RACE in Point of Service

AGAIN DIAMOND TIRES proved their **superior wear resistance** on all cars so equipped. Both regular and anti-skid Diamond Tires went the whole course of dry, grinding macadam roads without change.

They showed at the finish far less wear than any other make used.

THE DIAMOND RUBBER CO.
AKRON, OHIO
1876 Broadway, New York

MICHELIN WINS

AS USUAL

AT BRIARCLIFF

An Isotta car, driven by Lewis Strang, and equipped with flat and round Compressed Tread Michelin Tires and Michelin Demountable Rims, wins the greatest speed and endurance contest ever held in this country. This makes a clean sweep for Michelins in EVERY important contest here and abroad, not only this year but ever since motor racing began.

> The Briarcliff Trophy course is the most difficult and trying of any in this country for driving at speed. The numerous and very sharp turns, the varying character of the roadbed are comprehensive of all touring conditions and the course was selected for this reason. The distance 300 miles, was intended as a further endurance test. This victory of the Compressed Tread tires, a type exclusively Michelin's is a further proof, if any were needed, of the wonderful reliability, durability and consequent economy of Michelin Tires.

MOST RECENT VICTORIES OF THE MICHELIN COMPRESSED TREAD:

SAVANNAH—March 19th, '08, 342 mile race, Isotta Car, average speed 53.8 miles per hour; tires not touched from start to finish.

ORMOND—March 5th, '08, a New World's Record, Renault Car, 100 mile race. Average speed 82.17 miles per hour.

NEW YORK TO PARIS RACE—De Dion Car equipped with Michelins reached San Francisco, having had no tire trouble of any sort—front tires not reinflated between New York and 'Frisco.

MICHELIN TIRE CO.
MILLTOWN, N. J.

NEW YORK BRANCH - - - - 1763 BROADWAY

Photograph by Spooner & Wells.
AL. POOLE, IN THE ISOTTA, MAKING THE S. TURN AT TOP SPEED.

RENAULT

RENAULT REGULARITY

Has Again Been Confirmed by the Results of the

BRIARCLIFF RACE

In View of the Dangerous Course, the Renault Drivers Were Instructed Not to Take Chances. However, from Start to Finish They Improved Their Positions, Owing to the Regularity of Their Motors.

Two Renaults Started; Two Finished, Their Time Being:

		1st Lap.	2d Lap.	3d Lap.	4th Lap.	5th Lap.	6th Lap	7th Lap.
BLOCH	No. 18	48 min.	47 min.	46 min.	47 min.	50 min.	47 min.	47 min.
BERNIN	No. 14	59 min.	50 min.	47 min.	47 min.	50 min.	47 min.	47 min.

At the start Bernin loss 10 minutes fixing a loose magneto. At the fifth lap both drivers stopped 3 minutes to fill gasolene tanks.

Had the race been run in ten laps, as originally intended, results might have been different.

RENAULT CARS ARE BUILT FOR DURABILITY

A 35-45 H. P. REGULAR STOCK RENAULT HOLDS 24-hour WORLD'S RECORD

1079 Miles—An Average of 45 Miles per Hour for 24 Hours.

Renault Cars Are Guaranteed to Make the Trip Between New York and Chicago

ASK FOR A DEMONSTRATION

Renault-Frères Selling Branch, Broadway & 57th St., N.Y.

PAUL LACROIX, General Manager.   'Phone, 3004—Columbus

MR. G. D. HOLMES DRIVING HIS MIDLAND, A CAR NEW TO THE EAST.

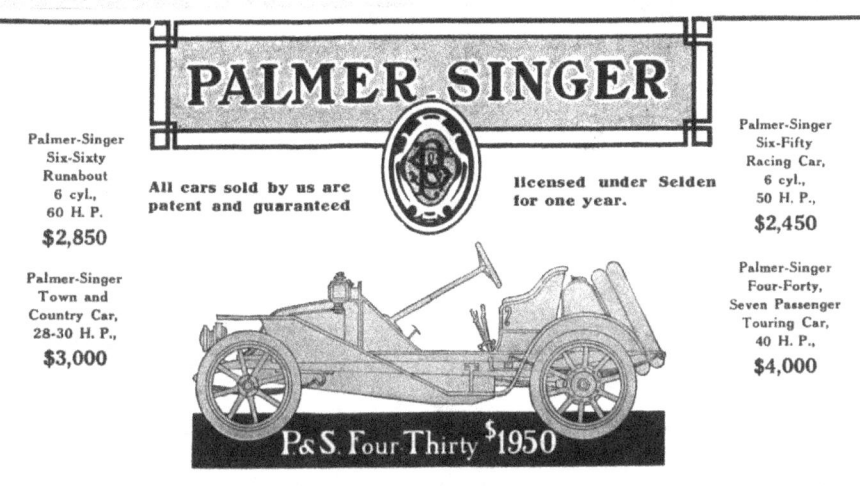

PALMER-SINGER

| Palmer-Singer Six-Sixty Runabout 6 cyl., 60 H. P. $2,850 | Palmer-Singer Six-Fifty Racing Car, 6 cyl., 50 H. P., $2,450 |

All cars sold by us are patent and guaranteed

licensed under Selden for one year.

Palmer-Singer Town and Country Car, 28-30 H. P., $3,000

Palmer-Singer Four-Forty, Seven Passenger Touring Car, 40 H. P., $4,000

P.&S. Four Thirty $1950

THE "Skimabout" is the best runabout in the world, bar none, for city use—and far better than most for country use. For the man who has little chance to go touring far afield it fills a long felt need.

The Palmer-Singer "Skimabout" is the most graceful little car on the market. It has all the lines of a foreign car, fair power, good speed, superb finish and style, and the best materials and workmanship that money can buy.

Metropolitan Distributors THE SELDEN

PALMER & SINGER MFG. CO.
1620-22-24 Broadway, NEW YORK
1321 Michigan Ave., CHICAGO

Sole Distributors THE SIMPLEX

HOLD THIS PICTURE AT ARM'S LENGTH.   THE MOUNTAINS HAVE A MESSAGE FOR YOU.

MAJA (pronounced "My-yah"), the climax of automobile construction, the masterpiece of the master makers, the perfected product of the famous Daimler Motoren Gesellschaft (Wiener-Neustadt), is their only production sold by the makers themselves to the American purchaser from THEIR OWN OFFICIAL BRANCH, through which genuine direct marketing the prices are lower than those of foreign cars of lesser fame.

IF YOU ARE GOING ABROAD and contemplate buying a car, you can rent at regular rates a brand new '08 Maja, body and equipment made to your order; and after your tour on deciding you want the car, have a liberal portion, of your rental charges credited toward its purchase, also saving much of the import duties on bringing it over. This is the most advantageous offer ever made, especially as you rent from a responsible house with every interest bound up in giving you perfect service, equipped for the purpose as is no other organization in the world.   We refer to many tourists who have adopted this method as to its satisfactory features.

AMERICAN BRANCH MAJA CO., Ltd. 58th St., Just East of Broadway, NEW YORK
Offices in NEW YORK, LONDON, PARIS, STUTTGART, HAMBURG and ST. PETERSBURG

Maja Foreign Touring Service has its depots in the above cities and correspondents throughout the world, arranges all licenses, fees, customs, insurance and shipping formalities, and does everything for the auto tourist irrespective of car used. Saves all the trouble. Write for information. Maja Continental Touring Guide and Map the only complete source of information printed in English for those who tour abroad. Guide $3. Map $2. Subscriptions now being received. Maja Map of Westchester Co. The most practical ever produced. Shows all roads, hotels, repair-shops, garages, country clubs, etc., police-traps, grades, danger spots, character of roads and every detail the motorist wants. On cloth, $1.50.

1909 — POPE=HARTFORD — 1909

$2750.00 : POPE-HARTFORD MODEL S PONY TONNEAU : $2750.00
Three Models: Touring Ca., Roadster and Pony Tonneau

CAMPBELL=CORWIN CO., MONTAUK GARAGE
910 Union Street, Brooklyn, N. Y.
EXCLUSIVE AGENTS FOR BROOKLYN AND LONG ISLAND

MERCANTILE MOTOR CAR CO. (Inc.)

Best Service Guaranteed

General Machinists

FIRST-CLASS CARS
Fully Equipped for Renting

Renting Station and Garage
140 and 142 Livingston St.

Telephone 1600 Main

Tel. 2538 Prospect

Automobiles Stored

And Repaired

Dealers in Second-hand Automobiles

Repair Shop, 368 Cumberland Street,   Brooklyn, N. Y.

Photographed for Brooklyn Life by F. A. Walter.
MR. ARTHUR E. LETHBRIDGE DRIVING HIS 25 H.P. CADILLAC.

A LITTLE ACTION—AT BRIARCLIFF

IN A *Stearns* "GUY VAUGHAN" RUNABOUT

3 cars entered 3—cars finished—all three ready as before to do it again

BEST STOCK CAR OF THE WORLD

BEDFORD AVE., 1287-91 Corner
ATLANTIC
PHONE, BEDFORD-4192

The Allen-Swan Co.

LEW H. ALLEN, Pres. and Gen'l Mgr.
HALSTEAD SWAN, Sec'y and Treas.
HOWARD DRAKELEY, Sales Dept.

THE INCOMPARABLE
WHITE
THE CAR FOR SERVICE

New Models of the White Steam Car

We will shortly offer a new car of 30 steam horse power which will be larger, roomier, stronger and far more speedy than any which we have previously made. In its construction it will follow the general lines of the well-known White System but with numerous important improvements suggested by the study and experience of the year. The various elements of the power plant—engine, generator, condenser, etc.—will be of increased dimensions, and every part of the car will be brought up to a new standard of strength, more than proportionate to the increase in power. This car to be known as the Model "G," will be fitted with two distinct styles of body: a Pullman body, seating seven, and a touring body, (illustrated above), seating five, and having most ample provisions for carrying baggage.

We will also offer a smaller car, to be known as the Model "H," which will closely resemble the present highly successful and popular Model "F," although the new car will have a somewhat shorter wheel-base. In the Model "H," which will be conservatively rated at 20 steam horse-power, will be incorporated a number of the improvements which will be found in our new Model "G."

A circular descriptive of the new models will be mailed on request.

WHITE SEWING MACHINE COMPANY
Cleveland, Ohio

We're always glad to answer inquiries as to correct livery usage.

To send samples and illustrations of correct liveries for every sort of man servant.

ROGERS, PEET & CO.

258—842—1260 Broadway

(3 Stores)

NEW YORK

50 H. P.	30 H. P.
TOLEDO	HARTFORD
$4,250	$2,750

Fully Guaranteed for One Year

From The
"New York Tribune,"
Sunday, Dec. 30, 1906.

The actual performance of a motor car on the road carries more weight with intending purchasers than volumes of stereotyped phrases. * * * The new 1907 model of the 50-horsepower **Pope-Toledo** touring car on the road last week showed a speed of 60 miles an hour, with its regular quota of passengers aboard. * * * The powerful **Pope-Toledo** quickly registered 42 miles an hour on third speed, and then gradually climbed to 45, and finally halted at 46. * * * Fourth speed was thrown in without a jar or noise, and at once the speedometer registered 54 miles an hour. Soon the 60-mile point was passed, and this high speed was held for some little distance. The car showed entire absence of vibration and held its course as straight as the flight of an arrow.

A. G. SOUTHWORTH CO.
INC.

1733 Broadway,   342 Flatbush Ave.,
New York.   Brooklyn.

We Exhibit at Madison Square Garden, January 12 to 19, 1907.

I. M. ALLEN GARAGE
116 SOUTH PORTLAND AVENUE

Bet. Fulton St. & Hanson Pl.   Telephone 4026 Prospect

NOW READY FOR...... **STORAGE**

Distributors of

STEVENS-DURYEA AUTOMOBILES

THE MATHESON
—SIGNIFIES—

That at last a Touring Car has been built in America which, for Speed, Power and Endurance, is unsurpassed.

The Prestige of This Country NOW Rests Upon

"THE SEVEN PASSENGER - SEVENTY MILES AN HOUR - - - **Touring Car**"

LICENSED UNDER SELDEN PATENT

IMMEDIATE DELIVERIES

THE MATHESON COMPANY OF NEW YORK
1619-1621 Broadway   Tel. 4876 Col.

We will exhibit only at the Seventh National Automobile Show, Madison Square Garden, Jan. 12-19, 1907.

The Brooklyn Home For

CARS

will be open on October 12th, at 42 Flatbush Avenue, extending through to 325 Livingston Street. ::   ::   ::

BUICK MOTOR COMPANY
FACTORIES at Flint, Michigan

Fastest Stock Car at Brighton Beach

The Stearns

Best STOCK Car of the World

Wins 50 Mile Stock Car, and 6 Cylinder 5 Mile Races

¶ On Friday afternoon the Stearns showed itself an easy winner against every other stock car, winning the five-mile race in 5 min. 20 2-5 sec., open to six-cylinder cars. In the 50-mile race the Stearns Six won in 50 min. 43 sec., a Stearns Four winning Second place. The winning car in each event was driven by Barney Oldfield.

STEARNS FASTEST IN 24-HOUR RACE

¶ Through an unfortunate accident due to an unavoidable collision the Stearns 24-Hour Car was so badly damaged that it was able to be on the track but 8 full hours out of the 24. During this time it showed itself to be

SWIFTER THAN ANY OF ITS COMPETITORS

¶ Its mileage per hour during those 8 hours was 50, 50, 53, 57, 52, 44, 46, 51. An average of 50 1-3 miles per hour. Barring accident this would have made the Stearns distance for the 24 hours, 1,208 miles—31 miles better than the winner.

STORY OF THE COLLISION

For some unaccountable reason Special Officer Fickert, who was guarding one of the gates leading to the paddock, stepped out on to the track. Marquis, driving the Stearns, was hugging the inside fence, and saw the officer when still some distance away. Marquis immediately guided his car toward the outside fence, but Robertson, driving the Simplex, which was slightly ahead of the Stearns, did not see the officer so quickly, and although he put on both brakes, it was impossible to avoid hitting the man on the track. The impact, together with the sudden setting of the brakes, threw the Simplex toward the outer rail, directly in front of the Stearns, which immediately crashed into the rear of Robertson's car, tearing loose the gas tank and tail lights. The crash smashed off the front end of the left Stearns frame ten inches back of the spring hanger, and tore a hole in the radiator. Much time was consumed in laboriously patching up the frame with temporary steel plates riveted to the side. At best this was only a makeshift, and the weakened parts soon separated again after a couple of hours. The Stearns car had absolutely no engine trouble of any kind, nor, barring stops for gasoline and oil, was it off the track for any other reason during the entire 24 hours, aside from tire replacements.

"The Stearns car had no mechanical trouble whatever throughout the entire race. The collision was a great disappointment to the thousands who were aware that with the Stearns car in the race, the finish would have been close and exciting."

"A LETTER FROM ROBERT LEE MORRELL:

New York City, October 5, 1908.

Mr. A. W. CHURCH, 1743 Broadway, New York City.

Dear Sir:—I want to express to you my disappointment at the unfortunate collision which put your Stearns car out of the running in the 24-hour race. No blame can be attached to the car with which you collided, as I understand the brakes had to be applied on the Simplex, which was directly in front of the Stearns, in the hope of saving the life of the officer who ventured on the track. It was, therefore, impossible for you to prevent hitting the car ahead.

Very truly yours,
(Signed) ROBERT LEE MORRELL."

The ALLEN=SWAN Co.

BROOKLYN

Phone Bedford 4200-01   Bedford Ave., 1287-91

LEW H. ALLEN, Pres. and Gen'l Mgr.   HALSTEAD SWAN, Sec'y and Treas.   HOWARD DRAKELEY, Sales Dept.

Here's the car
that has shaken the automobile world from centre to circumference

"Thirty"

Price, $1400.00 F. O. B. Detroit
(Including three oil lamps and horn)

JOS. D. ROURK
1001-3 Bedford Avenue

TELEPHONE: 3730 BEDFORD   OPEN EVENINGS

Locomobile

Our new 1909 Type I now ready for delivery. Demonstrating Car on hand. For further particulars write

The I. S. REMSON M'f'g Co.
Sole Agents for Brooklyn and Long Island
Garage—754-760 Bedford Avenue

HAYNES MODEL X 1909

The car of quality that has stood the test of time. Every necessary feature requisite to produce the perfect touring car is found in the model X for 1909. *Demonstrations.*

BOROUGH AUTOMOBILE CO.
679-681 McDonough Street
Bet. Howard and Saratoga Aves.
Renting, Repairing and Supplies

ELMORE

1909 Model 44

ELMORE, $2,500, Equipped

Bigger, faster, more powerful, more beautiful, more luxurious—the 1909 Elmore is the only car in the world at any price which surpasses the wonderful 1908 Elmore.

Take the most costly make of car in the world, one of the imported $8,000, 1909 model ones, totally eliminate its up-keep expense, and offer it to an Elmore owner in exchange for his 1908 Elmore. Would he make the trade? Not unless he could sell the imported car and get another Elmore. Why? The Elmore is **fool proof**—a novice **CAN'T HELP** but get ideal service from it. It is trouble proof—road troubles just don't happen to it. It will run farther in a **DAY** or a **YEAR** or in **FIVE YEARS** than cars of far greater speed, far greater power—three times the cost. When you pay the price asked for the other standard American makes you spend a sum which would buy enough **ELMORES** to last for 25 years.

The 1907 Elmore cars in the hands of hundreds of private owners far surpassed any factory output in the world in point of economy of service, the average 1907 Elmore going through an entire year at a repair and replacement cost of less than $5. This record caused the sale of every 1908 Elmore before June 1st of this year.

If you contemplate getting ANY car during 1909 you would better **LOOK** at the Elmore **NOW**. The 1908 output was sold out before June 1st. The 1909 Elmore is a bigger, better car—it will be sold out still earlier.

A. ELLIOT RANNEY CO.

Broadway at 61st St., New York

Photograph by N. Lazarnick.

FRANK LESCAULT AT WHEEL OF SIMPLEX.
After making the new world's record of 57 miles in one hour.

Photograph by N. Lazarnick.

GEORGE ROBERTSON AT WHEEL OF SIMPLEX.
Just after making new world's record of 1,177 miles in 24 hours

CONDUITE INTÉRIEURE 4 PLACES SUR CHASSIS 6 CV

Photographed for Brooklyn Life by F. A. Walter.
MR. S. EDWARD VERNON IN HIS NEW STEARNS CAR IN PROSPECT PARK.

Photographed for Brooklyn Life by F. A. Walter.
MR. ABRAHAM ABRAHAM'S NEW STEVENS-DURYEA IN FRONT OF HIS TOWN HOUSE.

The Motor Car of the Past Decade.

Three New Brooklyn Cars.

MR. EDWARD W. WALTON, OF 134 MONTAGUE STREET IN HIS POPE-HARTFORD.

MR. ARTHUR T. DAY DRIVING HIS 1909 ELMORE.

Photographed for Brooklyn Life by F. A. Walter.
DR. CHARLES E. MANNING, OF 480 PUTNAM AVENUE DRIVING HIS MITCHELL.

The MG Sports

THE STILES THREESOME.
TWO SEATER AND DICKEY SEAT, CONCEALED HOOD. SUITABLE FOR MOUNTING ON 12/70 M.G. MAGNA AND D2 TYPE MIDGET CHASSIS.

STILES LTD.,   3, BAKER STREET, W.1.

Every Inch a Car

40. H. P. "Forty" Semi-Touring (fully equipped) $1850

KISSELKAR

Low Up-Keep Combines With Road Ability, Riding Comfort and Smart Appearance

The trim, smart appearance, comfort and road ability of the KisselKar you can prove merely by riding in a KisselKar.

Now about low up-keep. A few hundred dollars built into a car makes a big difference in the service you will get out of it. No KisselKar is built with a light, overpowered chassis, as in some automobiles, to provide a low selling price. If you buy a KisselKar "Thirty" or any other model you get you get a car that has the structural strength according to its horsepower, which means the best of road service and lowest up-keep.

The wheel base on every model is liberal, as it must be, to have a roomy tonneau and to give the car the perfect general "balance" which results in greater riding comfort and low running cost.

A careful investigation will show you that for purchase price, economy, comfort, appearance and low running and up-keep cost, the KisselKar offers you the best values in America.

"Thirty," $1500 — "Forty," $1850 — "Fifty," $2350 — 60 H. P. "Six," $3000. 1½ to 2 Ton, 3, 4, 5 Ton Trucks, Delivery Wagons, etc.

Except the "Thirty," which has more than "regular" equipment, KisselKar prices include full equipment—speedometer, demountable rims, glass front, shock absorbers, lamps, etc.—everything belonging to a perfectly equipped and dressed car—nothing left to be bought separately.

1912 KISSELKAR PORTFOLIO
One of the Most Informing Automobile Books Ever Published

This book is in great demand because it takes up the automobile subject in a broader, more detailed way, than is usual. It gives a new standard by which to judge pleasure car values. Free on request.

Kissell Motor Car Co., 206 Kissel Ave., Hartford, Wis.

Branches and distributors in Boston, New York, Philadelphia, Cleveland, St. Louis, Dallas, Chicago, Milwaukee, Kansas City, Omaha, Denver, San Francisco, Los Angeles, Seattle, Portland, Minneapolis, and all other principal points throughout the United States.

A car of such manifest and extraordinary excellence—a chassis so costly—that it will upset all your previous notions of which is really the finest car made in America. We urge upon you nothing but this—**ride in the Craig-Toledo.** We will abide by the results.

THE CRAIG=TOLEDO MOTOR COMPANY
TOLEDO, OHIO

1907 "AMERICAN" ROADSTER

"NO NOISE BUT THE WIND"

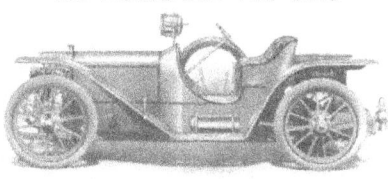

PRICE $3250. F.O.B. INDIANAPOLIS

WANT more than your money's worth? Most people do if they can *get* it. You can *get* it if you are a judge of Touring Car or Roadster *value*. If not, get an expert to investigate the 1907 American Roadster or American Tourist for you. His decision will suit us and *you*. At the price, no car at present approaches either. Above the price none are *better value*!

Experts admit that the 1907 American Roadster is in a class by itself. Look at the cuts of them. Look good—don't they? Imagine what either does with a 5x5 four-cylinder motor (water-cooled cylinders, offset valves on one side, operated by one cam shaft, served by an automatic carburetor with water-jacketed throttle designed especially for it, ignited by Bosch high-tension magneto) capable of turning up on the block 3000 revolutions, coupled to it. We *guarantee* 40 B. H. P. at 1000 revolutions! Both have floating type rear axle, nickel steel one piece front axle, frames and gears, ball bearing throughout except engine. Tie rod on Touring Car back of front axle. 20 inch flywheel, fan bladed, double reduction clutch, both models. No clutch slippage! No vibration! No skidding! Increased tire life, safety and comfort.

Every expert investigation means a sale. Every sale makes another. Every owner *roots* for these two honest American Cars because he gets *more* than his money's worth! Inquire and we'll prove it to *you*.

MEMBERS THE A. M. C. M. A.

"NO NOISE BUT THE WIND"

Some Excellent
Territory Unplaced

Cars selling fast
"GET BUSY."

PRICE $3250. F.O.B. INDIANAPOLIS.

1907 "AMERICAN" TOURIST

HAYNES 1921 CLOSED CARS

Utmost in beauty, luxury and utility—$1,000 underpriced

NOW, when the buyer at last is asking: "What am I getting for what I pay?" the advantage of the Haynes selling policy becomes increasingly evident. Enthusiastic Haynes owners have told us all along that the Haynes is $1,000 underpriced. The Haynes principle of building for the future has held good. We have been and are satisfied to produce the choicest car of its class and sell it at a price that is fair to the buyer and to ourselves.

The seven-passenger Haynes Suburban and the five-passenger Haynes Brougham richly deserve the high praise accorded them. Among closed cars they establish a class of their own. Quietly rich in finish and fittings, as such cars should be, they are distinguished in line and completely desirable in appearance. They are far and away beyond anything to be expected in their price-class.

A detailed description of the many superiorities of construction and design and of the thoughtful conveniences installed in each car is obviously impossible here. A personal inspection of these closed cars is invited and urged. To secure prompt delivery an immediate reservation is recommended.

THE HAYNES AUTOMOBILE COMPANY
KOKOMO, INDIANA   U.S.A.
Export Office: 1715 Broadway, New York City, U.S.A.

HAYNES CHARACTER CARS
Beauty — Strength — Power — Comfort

1893 — THE HAYNES IS AMERICA'S FIRST CAR — 1920

MAKING IT HOT IN A PLEASANT WAY

Gerd H. Henjes, vice president of Henry Henjes (the man with the gray overcoat), supervises the distribution of coal from his yards, which is given to needy who apply through the Home Talk-Bay Ridge Civic Council Emergency Relief Committee. Aside from the few bags distributed at the Henjes yard, 40 bags each week will be given unemployed at the committee's headquarters, 6756 4th Ave., where applicants must receive tickets from Mrs. C. A. Weaver, who is in charge of the office.

"ON THE MINUTE"
The finest limousines in service for Weddings, Dances, Parties, Theatres, etc.
THREE-STAR CABS
Phone PROSPECT-243

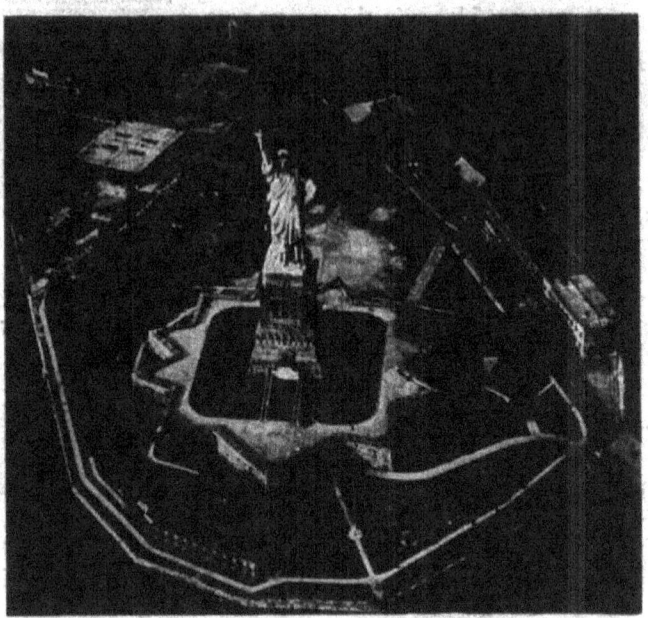

"The Statue of Liberty, in New York Harbor"

The sign of a reliable dealer, and the world's best Gasoline

THE manufacturers of Socony Gasoline were already refining petroleum when the French government presented the Statue of Liberty to this country in 1876.

Socony Gasoline is straight-distilled, pure and always uniform wherever and whenever you buy it. Its consistent quality means much in satisfaction and economy.

STANDARD OIL CO. OF NEW YORK

SOCONY
REG. U.S. PAT. OFF.
MOTOR GASOLINE
"Every Gallon the Same"

1933 Automobile Show Opens At the Grand Central Palace

The automobile makers of the nation today set the latest 1933 models before the public at the opening of the 33d National Automobile Show at the Grand Central Palace, Manhattan.

Opening this afternoon, 36 makes of motor vehicles, including nine truck models and one foreign make passenger car, ranging in price from around $330 to many thousands of dollars, are on display. The value of the exhibit exceeds $1,000,000.

Stream lines, with longer and lower bodies, feature the new season's models. Much attention has been given to interiors with wider seats in many instances and new ideas in seat adjusting and ventilation.

The National Automobile Chamber of Commerce, under whose auspices the show is given, has given particular attention this year to providing a luxurious setting for the cars. The Palace has been transformed by the use of silken fabrics, mirrored pillars and murals into a sparkling display center. Manufacturers also have outdone their efforts of other years in providing settings for their models with many special exhibits of the workings of the new devices incorporated in their cars.

The show, which will last a week, will be closed tomorrow, but will reopen Monday. The doors will open at 10 a. m. and close at 10:30 p.m. Many manufacturers are conducting separate exhibits at hotels and showrooms during the week.

BROOKLYN'S USED CAR MARKET PLACE

Automobiles for Sale 11

BUICK 1941 2-door special sedan, heater, slip covers, low mileage, perfect condition; make offer. MAin 2-0247.

BUICK '41 4" 4-Door Trk. Sed. $1,050
KINGS COUNTY BUICK INC
4th Ave. & 46th St. SHore Road 5-9696

CHEVROLET 1941 4-door sport sedan, radio, whitewalls, perfect, low mileage, original owner. No dealers. MAnsfield 6-8987.

CHEVROLET 1940 master de luxe sedan, like new, used room and demonstration $595. Etna Warehouse, 493 Monroe St. (Sumner).

CHRYSLER 1936 8-cylinder business coupe, heater, tires and mechanical condition good, sacrifice private. Dickens 2-3643.

DE SOTO 1941 fluid drive sedan, low mileage, fully equipped, cost $1,350, price $850. Box R 40, Eagle.

DODGE sport coupe 1937, good condition, heater, new battery, 5 good tires, low mileage, no reasonable offer refused. SHore Road 5-5406.

FORD 1936 Tudor, $50 cash or price, 28 7th Ave., 10 a.m. to 2 p.m. daily.

FORD 1938 de luxe 4-door sedan, good condition, $200, private owner. Christiansen. WIndsor 9-5959.

FORD 1938 2-DOOR, GOOD CONDITION; RADIO, HEATER. NA. 8-1253 (PRIVATE).

FORD 1941 2-door sedan with radio $112 cash; $27 monthly for 21 months. HOlls 5-5042-W.

FORD coupe 1935, excellent appearance and condition, $60. After 6 p.m. Price, 374 Prospect Place.

LINCOLN '37 7-Pass. Limousine, Bargain
J. J. HART, INC. FORD OF BROOKLYN
Bedford Ave. cor. Fulton St. MAin 2-2857

LINCOLN ZEPHYR, excellent condition, rubber perfect, new paint job, radio. 462 39th St. evenings.

NASH (Lafayette) 1937 2-door sedan, radio, tires practically new, good condition, sacrifice. Tadier, 129 Woodbine St.

Packard '40 "1g. Sedan $875
PACKARD MOTOR CAR CO OF N. Y.
1050 Atlantic Ave. (Classon). MAin 2-3400

PACKARD 1939 120 8-cylinder coupe sedan, radio and heater, 15,000 miles, excellent condition $650. SOuth 8-4549.

PLYMOUTH 37 4-DOOR SEDAN, RADIO HEATER, GOOD CONDITION, PRIVATE. $265. NEVINS 8-1007.

PLYMOUTH 1936 2-door coupe, rumble seat, heater, radio, 7 tires, good condition; reasonable. SOuth 8-9128.

Automobiles for Sale 11

PONTIAC 1940 4-door sedan, radio, heater, new rubber, fine condition, $750. 259 93d St. SHore Road 8-6416.

Credit Slips

CREDIT SLIP $300 for new Ford very reasonable. Call evenings SHore Road 8-1231.

Trucks & Commercial Cars 11b

FORD DUMP TRUCK 1938 CHEAP
OWNER LEAVING TOWN.
NEVINS 8-3514

Auto Dead Storage—Garages 16

AUTO STORAGE $3 PER MONTH
Stocking and Battery Service
TERMINAL WAREHOUSE
961 Halsey St. (Broadway). GL. 5-3333

Wanted—Automobiles 17

CARS wanted: high cash prices paid for late models only. Leviek Bros. Inc. 1285 Bushwick Ave., Brooklyn. GLenmore 5-7174.

CARS WANTED—High prices paid; cash waiting, no dealers. Etna Warehouse, 493 Monroe St. (Sumner). JEfferson 3-8474.

Cars Wanted—Top Prices Paid

Pickin Auto 225 Penn'a Ave. AP. 7-0088.

CASH WAITING — All makes, models, no led ape 690 Fulton St. corner Lafayette Ave. and Fort Greene Place. STerling 3-9297.

LATE MODEL CAR wanted from private party; will pay immediate cash, no dealers. Mr. Raab, MAin 2-4130.

WANT TO SELL YOUR CAR...
Have you a car you are not going to use this Winter? If so, now is a splendid time to sell it. The demand is big. And prices are good, too. Call me. The number is MAin 4-6200. I will gladly tell you the probable current market value of your car and help you write your ad. No obligation. Ask for Mr. Frankfort, Eagle Used Car Manager.

**BUY U. S. DEFENSE BONDS
AND SAVINGS STAMPS**

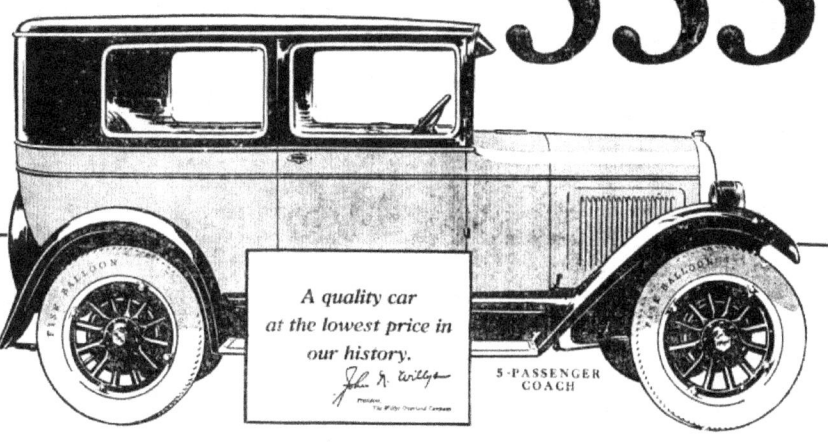

New Chevrolet Imperial Landau

AMUSEMENTS—MANHATTAN. AMUSEMENTS—MANHATTAN.

28th Annual NATIONAL **AUTO SHOW**

Opens TOMORROW, 2 P.M. Daily Thereafter (Except Sunday) 10 A.M. to 10:30 P.M.

JAN. 7 to 14

The Newest in Cars, Accessories and Light Trucks— Shop Equipment Section Open to Public After 5 P. M. Daily

Two entrances, Park Ave. and Lexington Ave.

GRAND CENTRAL PALACE

Adm. 75¢

AMUSEMENTS—MANHATTAN. AMUSEMENTS—BROOKLYN.

At the Show!

STUDEBAKER COMMANDER

World's Champion Car

$1495
f. o. b. factory

25,000 miles in less than 23,000 minutes —nothing else on earth ever ran so far so fast

See the identical car which made the above record—now on exhibit in the Studebaker Booth

AUTO SHOW

GRAND CENTRAL PALACE

JANUARY 7-14 10 A.M. TO 10.30 P.M.   OPENS TODAY AT 2 P.M

SUNDAY EXCEPTED

See What's New at This Great Show!

• CARS • ACCESSORIES • COMMERCIAL VEHICLES •

Hundreds of innovations for 1933. Never before such car values.   ADM. 75c

"AFTER-THE-SHOW" USED CAR SALE

Millions of people have visited the Automobile Shows all over the country — thousands have traded in their used cars against the newer, speedier, more beautiful models.

Dealers in Brooklyn, Manhattan, Long Island City and Jamaica have reconditioned these "trade-ins." They have made them mechanically perfect and almost new in appearance.

They are ready NOW — ready with a complete stock of reconditioned automobiles that will give long and satisfactory service — service backed by the guarantee of established automobile merchants.

Buy Them Now — the Prices Are Reasonable — the Cars Are Good.

Turn to
WANT AD SECTION
today and every day in the
BROOKLYN EAGLE

Below is the list of dealers who are offering unusual bargains during this stupendous AFTER-THE-SHOW USED CAR SALE:

AMERLING KAMNER CHEVROLET
1301 13th Avenue

BERRY BROTHERS
1265 Bedford Avenue

BISHOP McCORMICK & BISHOP
Atlantic and Bedford Avenues

BRIGHTON AUTO EXCHANGE
1077 Atlantic Avenue

BRAEMER AUTO SALES
1296 and 1110 Bedford Avenue

CADILLAC BROOKLYN BRANCH
749 Atlantic Avenue

CAMPBELL SALES & SERVICE
298 Pacific Street

CAPLAN, HENRY
310 Cushing Street

COLONIAL DISCOUNT CORP
205 Fulton Street

COLCORD—KELL CORP
1538 Bedford Avenue

DAVID CONDON, Inc.
60th Street and Fort Hamilton Pkwy

ETNA WAREHOUSE
105 Monroe Street

FRANKLIN FACTORY BRANCH
1807 Broadway, New York City

FRANKLIN-PASE CO
1900 Atlantic Avenue

GALLAGHER'S AUTO SALES
79th Street and Fourth Avenue

J. J. HART, Inc
1035 Atlantic Avenue

HETZER CHEVROLET
198 Fourth Avenue

KINGS COUNTY BUICK, Inc
44 Empire Boulevard

JOHN KOLETTY
1117 Bedford Avenue

KRUSE MOTOR CAR CO.
1365 Flatbush Avenue

LINCOLN SALON
1042 Atlantic Avenue

LEVINE WM.
510 Coney Island Avenue

LYNAGH & MAGEE
936 Bergen Street

MILLIGAN AUTO CO
1622 Atlantic Avenue

MULLER AUTOS
1893 Atlantic Avenue

NEUSTEIN'S GARAGE
111 E. 11th St. N. Y. C.

PACKARD MOTOR CAR CO.
Atlantic and Classon Avenues

PIERCE-ARROW MOTOR CAR CO
Atlantic and Grand Avenues

REO MOTOR CAR CO.
1530 Bedford Avenue

SANDMAN CHRYSLER PLYMOUTH
242 Flatbush Avenue

SHORE ROAD MOTORS
3924 Fourth Avenue

SIMONS STEWART CO., Inc
1491 Bedford Avenue

STUDEBAKER CORP
1353 Flatbush Avenue

VON KAMPEN MOTOR
1313 Rogers Avenue

WILLIAMSBURG AUTO
450 Clermont Avenue

Now Ready

First Showing January 1st and 2nd

–a New ESSEX SUPER SIX

New Bodies - Larger and Roomier –
New Appearance from Radiator to Tail Light –
Finer Fittings - Four Wheel Brakes –
High Compression - Long Life Motor &
An Amazing Price

Beautiful *from* every angle . .

THE SEDAN, in two shades of blue, with cream striping, is larger and roomier, with lines fitting seats, wider doors, rich upholstery and appointments.
$795
f. o. b. Detroit, plus war excise tax

THE COUPE has wide seat, ample luggage space in the rear deck, and a comfortable leather rumble seat which is removable.
$775
f. o. b. Detroit, plus war excise tax

THE COACH is larger, wider, roomier—a five passenger Superbox, as distinctive in appearance as it is practical.
$735
f. o. b. Detroit, plus war excise tax

One look at the new Essex Super-Six will convince you that it will excel in popularity the Essex which has just completed the most successful year ever achieved by a six-cylinder car.

From radiator to tail light it is a smarter, more beautiful car than even the Essex which preceded it. And in performance it surpasses in smoothness, reliability, speed and ease of handling, the standard Essex owners are so proud to acclaim.

You get an impression of completeness and fine quality in the design of every detail. From the new pattern Colonial lights—the higher, narrower radiator with vertical shutters—the heavy sweeping fenders—the rubber-covered running boards—the new improved steering mechanism and the steering wheel similar in design and construction to that now used on the very latest and very highest priced cars—there is outstanding reason for pride.

The bodies are not only new and roomier but are so designed as to give a lasting, solid, rigid service. The roof is flatter—the car is bigger in fact and in appearance. Door fittings, hinges and locks are impressively substantial and lastingly beautiful.

The upholstering is not only durable but is also rich to eye and touch. The seats are form fitting—the backs high and comfortable, the leg room for driver and passengers is generous.

But rich and inspiring as is the appearance of this new Essex, still Essex surpasses itself in performance.

Its universally acknowledged supremacy in get-away and its ability to travel at top speed all day long reaches a new limit.

The get-away is perceptibly faster. You will find at the end of a day's run that you have covered more miles than was possible before.

Essex steering, long famous for its ease, is now smoother than you will find in most cars, regardless of their cost.

The Bendix four-wheel brakes give complete and attention-free control of your car at all speeds over every road condition.

No car near its price uses such large tires—30x5 inches. They may be driven with less air pressure and, of course, add hundreds of miles to tire life.

You can't help feeling that in the new Essex, quality and finest detail are outstanding. It will impress you as being much more than a serviceable transportation vehicle, for in the beauty of its lines, the smartness of every detail, the character of its richly lacquered bodies, the softness of its seats, the feel of its upholstering, as well as its performance, there is everywhere cause for your admiration.

The world's largest sale of six-cylinder cars became an Essex achievement solely because of merit. The car we now invite you to see is so outstandingly superior to anything you can have imagined that you must expect it to command a higher price.

But with all these advantages, there is also an amazing price reduction. The Sedan at $795 f.o.b. Detroit is $40 below the Sedan price of last year.

Buyers can pay for cars out of income at lowest available charge for interest, handling and insurance

HUDSON MOTOR CAR COMPANY OF NEW YORK, Inc.
BROOKLYN, N. Y.—1422 Bedford Avenue   JAMAICA, N. Y.—Bergen and Hillside Avenues

BROOKLYN   **QUEENS**

Bensonhurst Hudson-Essex Co., 86th St. and 18th Ave.
Evergreen Motor Sales, Inc., 1519 Bushwick Ave.
Greenpoint Hudson-Essex, 576-580 Manhattan Ave.
Lenox Motor Sales Co., 837 Roebling St.
Bedford Hudson-Essex Co., Inc., 1194 Bedford Ave.
Flatbush Hudson-Essex Co., Inc., Seventh and Flatbush Aves.
Benj. F. Stephens, Flatbush and Bedford Aves.
Parkway Hudson-Essex Co., Inc., 4308 Fort Hamilton P'kway.

Corona Hudson-Essex Co., Inc., Northern Blvd. and 99th St., Corona, L. I.
Queens Hudson-Essex, Inc., 215-03 Jamaica Ave., Queens Village, L. I.
Forest Hills Hudson-Essex, Inc., 118-28 Queens Blvd., Forest Hills, L. I.
Flushing Auto Sales Corp., 18 Farrington St., Flushing, L. I.
Shedlin Brothers, 1414 Central Ave., Far Rockaway, L. I.
Schwind & Parker, 2d and Newtown Aves., L. I. City, L. I.
Hoffman Sales & Service, 2820 Cooper Ave., Glendale, L. I.

Details You Will Note

New size—larger, longer, wider, inside and out.

Higher radiator with vertical lacquered radiator shutters—on no other car under $2,000, Hudson excepted.

Wider, heavier fenders, not found in this price class.

Colonial type headlamps and saddle type side lamps.

Bendix four-wheel brakes, the type used on the most expensive cars.

Silenced body construction, reinforced, rigid and durable.

Five-inch tires, a full size larger than used on any other car of this weight and price.

Wider doors, for easy entry and exit.

Worm and tooth disc design steering mechanism, used only by costly cars.

Electro-lock type of theft protection used in high-priced cars.

Adjustable tire carriers (for fitting with or without trunk).

Fine grade patterned velour upholstery.

Wider, higher, form-fitting seats.

New instrument board, finished in polished ebony, grouping motometer, ammeter, speedometer, gasoline and oil gauges.

Starter on instrument board, quick, convenient, positive.

Steering wheel of black hard rubber with steel core, and finger scalloped, a detail of costly car appointment.

Light, horn and throttle controls on steering wheel.

Rubber-matted running board.

PRIZE WINNER ACCEPTING PONTIAC

The above photo shows F. A. Gehrhardt, sales manager of Kings County Buick, Inc., Buick and Pontiac distributors for this borough, presenting Herbert Harris and his family of 226 Skillman St. with a new 1933 Pontiac Straight 8 Sedan, first prize in the recent radio contest. The presentation took place at the headquarters of Kings County Buick, Inc., 44 Empire Boulevard.

AUTOMOBILE DEAD STORAGE

$6 to $8 Per Month

The Brooklyn Warehouse & Storage Co.
335-355 Schermerhorn St.
Brooklyn, New York

Phone Cumberland 0200

Associated Press photos
Armed Swiss troops moving on crowds

A Greater Margin of Safety—the Holland Way

The up-to-date equipment and modern methods used in the Holland Laundry are best because they are safest.

In the Third Annual Accident Prevention Campaign, sponsored by the Associated Industries of New York State, and covering a period of thirteen weeks, the Holland Laundry proved its efficiency and carried off the highest honors by maintaining a spotless record.

To Holland employees there was nothing unusual in that. The equipment they operate and the methods they use are the best that money can buy and man devise for preserving a greater margin of safety, economy and efficiency in laundering your linen.

The housewife who has her laundry done the Holland way entrusts it to men who know their business, and is assured of the best results possible to obtain—and at no greater cost.

Careful handling throughout, perfect cleanliness and prompt deliveries are synonymous with the Holland way. When viewed from any angle it provides the greatest margin of safety.

ALL YOUR FAMILY LAUNDERING

Men's shirts and collars, women's and children's wearing apparel, handkerchiefs, towels, bed and table linen — all included and returned perfectly ironed ready to wear, for 18 CENTS PER POUND. Minimum 10 pounds.

18¢ Pound

NO EXTRA CHARGES of any kind on bundles at least half bed and table linen.

NO MARKING except on men's shirts and collars.

DELIVERED IN THREE DAYS after collection, packed in a neat, strong box for protection in transit.

35 Modern Electric Delivery Trucks to Serve You

HOLLAND LAUNDRY, Inc.

225 to 235 Twenty-fifth Street
Brooklyn, N. Y.
Phone Huguenot 1800

Greater even than its beauty is the performance of the new Ford car

MILLIONS of people have seen the new Ford since it was first announced on December 2nd and have been delighted with its smart low lines, its sturdy rugged strength, and its beautiful colors.

The art of the master designer is evident not only in the graceful contour of radiator, body and fenders, but in the harmonious relation of all features, so that the car as a whole is extremely pleasing to the eye.

In every least little detail, your impression of the new Ford is one of substantial simplicity and richness — a car that is entirely new and modern, yet with a quiet style that is always in good taste in any company.

Motor car beauty of a new and unusual kind is indeed revealed in the new Ford. Yet this beauty, striking though it is, is but one of the many features of this new car.

Your greatest thrill will come when you are set behind the wheel of the new Ford and know the thrill of driving it. Then you will have a full appreciation of what this car can do. Then you will know that it is not just a new automobile — not just a new model — but the advanced expression of a wholly new idea in modern, economical transportation.

For here is the complete car. Here, at a low price, is everything you want or need in a modern automobile . . . speed of 55 to 65 miles an hour . . . 40 horse-power engine . . . acceleration from 5 to 25 miles an hour in 8½ seconds in tests with a Tudor Sedan body and two passengers, and even quicker acceleration in the Roadster, Coupe and Sport Coupe . . . exceptional hill-climbing qualities . . . 20 to 30 miles per gallon of gasoline, depending on your speed . . . four-wheel brakes . . . Houdaille hydraulic shock absorbers . . . easy-riding transverse, semi-elliptic springs . . . typical Ford reliability and low up-keep cost . . . Even a Triplex shatter-proof glass windshield is given you in the new Ford without extra cost.

The outstanding performance of the new Ford is the direct result of the quality that has been built into every inch of it.

Its beauty is not confined to externals only, but goes deep down into every part of the car — even to those hidden, covered parts which you may never see.

Throughout, the new Ford is an example of fine automobile engineering. Its sound mechanical beauty delights the engineer and technical man, even as its unusual beauty of line and color delights the artist. Many features of it are exclusive Ford developments. Some are wholly new in automobile practice.

No use to say to you — make it a point to see the new Ford and arrange for a demonstration as soon as possible.

By its performance you will know that it is the most unusual value ever offered in a motor car. By its performance you will know that there is nothing quite like it anywhere in design, quality and price.

The new Ford Roadster sells for $385; the Phaeton for $395; the Tudor Sedan for $495; the Coupe for $495; the Sport Coupe with rumble seat for $550; and the Fordor Sedan for $570. (All prices are F. O. B. Detroit.)

Standard equipment includes five steel-spoke wheels, four 30x4.50 balloon tires, windshield wiper, speedometer, gasoline gauge on the instrument panel, dash light, mirror, combination stop and tail light, oil measuring rod, complete tool equipment, theft-proof coincidental lock, pressure grease gun lubrication, and Triplex shatter-proof glass windshield.

FORD MOTOR COMPANY
Detroit, Michigan

New STAR Four
A Durant Product

LARGEST CAR AT ITS PRICE
LOWEST PRICED CAR OF ITS SIZE

Sport Roadster $495   Two-Door Sedan $495
Coupe . . . . $495   Four-Door Sedan $570

Prices f. o. b. Lansing, Mich.

SPECIFICATIONS

107 inch wheelbase   Full crown fenders
Four wheel brakes   Long flexible springs
Streamline body, wide doors   Gas tank in rear
Rubber mounted motor   Choice of four colors

2-Door Sedan
$495
f. o. b. Lansing, Mich.

See It at the Show!

DURANT MOTORS, INC., New York, N. Y.

Factories: ELIZABETH, N. J.   LANSING, MICH.   OAKLAND, CALIF.   LEASIDE, ONT., CANADA

DURANT

AT THE NEW YORK SHOW YOU WILL SEE W. C. DURANT'S GREATEST CONTRIBUTION TO THE AUTOMOTIVE INDUSTRY · · A COMPLETE LINE OF SIXES BEARING HIS NAME, IN THREE DISTINCT SERIES

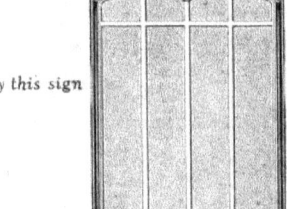

"By this sign   shall ye know it"

DURANT MOTORS Inc., NEW YORK
Factories: ELIZABETH, N. J. LANSING, MICH. OAKLAND, CALIF. LEASIDE, ONT., CANADA

DURANT

"I've seen all the new low priced cars and ROCKNE'S NUMBER ONE IN QUALITY!"

ONLY ROCKNE HAS THEM ALL!

YOU won't have to wait for the end of 1933 to find out which car is the low-priced field's Number One Car. That's already settled . . . *by Rockne!* In appearance, performance, comfort, quality—*and particularly in character*—Rockne is 'way out in front.

Furthermore, this roomy, luxurious 1933 Rockne is far ahead in every way of the Rockne that went from 31st place in sales last January to 8th place in 8 months . . . the Rockne that had the biggest per cent of the total volume ever sold by a first year car!

This 1933 Rockne hasn't been cheapened or skimped . . . in fact, it reaches into the higher priced field for its specifications.

No car that sells within $250 above the 1933 Rockne's price has such luxurious interiors, such riding comfort. It's speedy, it's safe, it's smartly superior in style. It's a dream of a car to drive—a positive marvel of economy to maintain. Come in today and arrange to take out a 1933 Rockne for a convincing trial drive!

Six cylinder 70 h. p. engine floated in live rubber! • Free wheeling, synchronized shift and silent second! • Automatic switch key starting! • Double-drop, rigid X-frame! • Silent threaded spring shackles! • Quadruply counter-weighted crankshafts! • Electro-plated pistons! • Hydraulic shock absorbers! Extra large brakes! • Extra large capacity batteries! • Safety glass windshields! • Closed bodies wired for radio! • Extra roomy, more luxurious interiors! • Contoured upholstery with special coil springs! • One-piece, all-steel bodies! • Full aerodynamic lines! Smaller wheels, lower over-all height!

Space B-4, Second Floor, at the Show

The Studebaker Sales Corporation of America

There is a Salesroom Conveniently Located in Your Neighborhood

$585
and up at the factory

NEW 1933 ROCKNE SIX
SPONSORED AND GUARANTEED BY STUDEBAKER

33d Automobile Show Opens to Big Crowd

Increasing Degree of Beauty in Models Exhibited Streamline Effect Prevails—Hood-Fender Unification New Feature

By JOHN J. A. O'NEILL
Science Editor of The Eagle

The Thirty-third National Automobile Show which opened yesterday afternoon at Grand Central Palace is an exhibition of the progress that ensues when scientists, engineers, artists and businessmen join forces to excel themselves in producing motor car passenger transportation.

An increasing degree of beauty in exterior design, well exemplified at this year's show, indicates that the artistic forces are gaining increasing recognition from the production departments of the automobile manufacturers, while the increasing simplification of operating controls and efficiency in power plant indicate the engineering staffs are busily engaged between shows.

Bodies More Streamlined

Cars this year are more curvilinear, a further development of the trend that was well under way last year. The bodies are becoming more and more streamlined. Practically all cars shown have the sloping front on the radiator shield and more graceful lines to the fenders, particularly in front, producing an effect which merges the hood, fenders and radiator shield into a harmonies unit instead of an assembly of mechanical parts.

Sloping windshields, curved body tops and curved backs that drop gracefully to the beaver tail outlines of the rear ends are to be found on almost all the passenger cars exhibited.

These features of design are found exemplified to the greatest extent in the "tear drop" model of the Pierce-Arrow. A head-on view of this car presents a striking appearance. Unification of the hood, radiator shell and fenders has been carried to such an extreme that it is difficult to identify any of these separate parts in the assembly. If the lines were somewhat more extreme the car would strongly resemble an elongated egg. As it is the interior of the car has distinct egg-shaped outlines. This is a custom model.

Hood Fender Unification

Among the more moderately priced cars the Willys-Knight carries this hood-fender unification scheme to the greatest extreme, its headlights being made an integral part of the elaborated fenders.

Practically every one of lines of this year's displays the advanced movement in body design that was anticipated to the greatest extent at last year's show by the Graham-Paige, with its flaring fenders and sloping front, and which was given support by the curved radiator shield of the De Soto. The other curved body lines so common this year have been borrowed from the custom jobs at recent salons. As a result the stock models of this year's cars are from an artistic point of view competitors of recent salon models.

29 Makers Exhibiting

Twenty-nine makers of passenger cars are exhibiting at the Palace. There are also six truck manufacturers' exhibits and many accessory exhibits. The floor space occupied is the same as last year.

Passenger cars exhibited are Auburn, Austin, Buick, Cadillac, Chevrolet, Chrysler, Continental, De Soto, Dodge, Dubonnet, Essex, Franklin, Graham, Hudson, Hupmobile, La Salle, Lincoln, Marmon, Nash, Oldsmobile, Packard, Pierce-Arrow, Plymouth, Pontiac, Reo, Rockne, Studebaker, Stutz and Willys.

More attention is being concentrated in the exhibits in the complete ensemble than in the mechanical details, but numerous innovations in the latter are to be seen.

French Innovation Shown

One of the most unusual mechanical developments is that exhibited in the Dubonnet car. This car is not in production but was built in France to exhibit some innovations which its producers hope will be adopted by American car manufacturers. Chief among them is the mechanical suspension of the wheels. Each wheel is individually mounted on the chassis through the suspension device, thus eliminating front and rear axles. The bottom of the quite thick chassis is but eight inches above the road. The gasoline tank is in the frame of the chassis and the drive shaft passes over it. The rear wheels are connected through universals to the drive in the differential housing.

The hall is beautifully decorated mirror effects being used generously. The opening attendance was large.

FRANKLYN BAUR
The famous Radio Tenor with his new Cadillac Town Sedan

ALL SET FOR A RIDE TO THE PICNIC GROUNDS

One of the features of the motor-car ball, held Friday night at the Commodore Hotel, in Manhattan, on the eve of the opening of the 1933 Automobile Show at Grand Central Palace, was the showing of how milady rode in the horseless carriage of 1903. In the car are Miss Alice L. Bliss of Bellport, L. I. (left), and Mrs. Theodora Fera, dressed in the daring costumes of the flapper of those days.

Announcing

The VICTORY SIX
BY DODGE BROTHERS

Simple fairness to this remarkable new product calls for words that would seem extravagant were they not so obviously and accurately truthful.

From an engineering standpoint The Victory is radically new, radically different and thoroughly original—is literally years ahead of its time in many vital features.

Subjected to long and peculiarly grueling tests over every kind of primitive road, it has emerged completely the victor—fit and ready to serve the advanced and exacting needs of today and tomorrow.

In a very real sense, too, it compares with no other car or class, because NO car, either here or abroad, provides features that are comparable.

To enjoy these advantages you must buy THIS car, for elsewhere they simply do not exist.

Revolutionary New Principles of Design

Two revolutionary new principles differentiate The Victory from all other motor cars:

1. For the first time since the invention of the automobile, the chassis and body of The Victory are a single integral unit—the wide, deep Victory chassis frame, flush with the lines of the body, replaces the customary body sills. (Heretofore, the body was mounted on a sill and both in turn mounted on the chassis.)

2. For the first time in history, battleship construction (i.e., double steel walls) is applied to the motor car.

The results of these, and other basic innovations are astonishing in their effect on every phase of motor car value, beauty, comfort, safety, strength and most impressive and important of all—*performance itself*.

Spectacular Performance

With chassis and body a single unit, there are 350 fewer parts—175 pounds less weight—and an extremely low center of gravity.

The results are greater motor efficiency—increased power in relation to load—quicker pick-up—greater stability and flexibility—an easier car to handle—a faster car to drive!

21 miles to the gallon at 25 miles per hour is precisely what you can expect—with sustained high speed all day long at instant call.

Drive over cobbles and await the usual discomfort—it will never occur! The Victory is the smoothest riding car, for its type, ever built.

Safe, Strong and Stable

Because the chassis frame conforms precisely with the body lines — *with no body overhang*—and because of the car's low center of gravity (weight close to the ground) The Victory is remarkably stable—tipping, skidding and swaying are reduced to a point positively negligible!

Turn a sharp corner and you will understand!

And the double steel walls mean double protection in case of accident—double the safety of any type yet known. A staunch body, with doors that close with a substantial and non-metallic snap.

Internal-expanding Lockheed-Hydraulic brakes, a rigid, 8-inch chassis frame, wide windows, full-vision windshield and thin steel corner posts are further vital factors of greater safety.

Quiet and Comfortable

The aim of all engineering is simplicity, for simplicity means economy—strength—SILENCE.

There are only 8 major parts in The Victory body—and they are welded into a single unit. Not a joint to squeak or rattle. Body and chassis act together, mutually flexing, mutually supporting, without stress or strain. A SILENT body.

And Dodge Brothers powerful new six-cylinder motor—specially engineered for The Victory—preserves this luxurious quietness at all speeds.

A Car of Striking Beauty

Body and chassis built as a single unit—without the customary body sill—permits lower over-all height with liberal head-room and road-clearance.

Splash shields, a constant source of noise and annoyance are replaced by the wide, deep Victory chassis frame; wide heavy-gauge one-piece fenders and drum-type head-lamps are provided. Upholstery, hardware, instruments, color combinations and other appointments satisfy the most exacting demands of style and good taste.

Indeed, you have a distinct and thrilling impression that the car in which you are riding is long, low, swift and safe—a car of surpassing originality and smartness down to the smallest detail.

And you are equally conscious that in the production of this car, Dodge Brothers have adhered strictly to the standards of dependability and long life which have distinguished their product for more than thirteen years.

Proceeding on this solid foundation, they have achieved a result as original and striking by comparison as was the first Dodge Brothers motor car displayed to the public on January 1st, 1915.

Tune in on WEAF for Dodge Brothers Radio Program every Thursday Night, 8 to 8:30—National Broadcasting Company Network

$1095
4-DOOR SEDAN, F. O. B. DETROIT

BISHOP, McCORMICK & BISHOP
NEW YORK — BROADWAY and 57th Street
BROOKLYN — 1221 Bedford Avenue

ON DISPLAY TODAY WITH THE SENIOR SIX AND AMERICA'S FASTEST FOUR
A Special Showing of Dodge Brothers Complete Line — The Palm Room, Hotel Pennsylvania — 1221 Bedford Avenue, Brooklyn — Broadway and 57th Street — January 7th-14th

Announcing the New Series
PONTIAC SIX
[With FOUR-WHEEL BRAKES]

— a Successful Six
NEW BIDS FOR EVEN GREATER SUCCESS

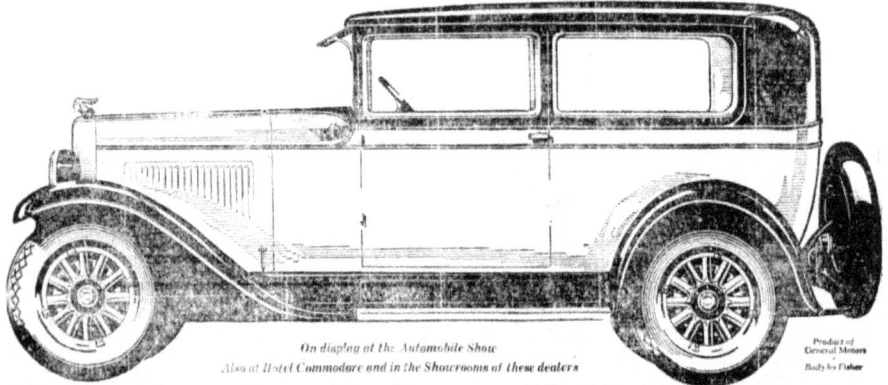

On display at the Automobile Show
Also at Hotel Commodore and in the Showrooms of these dealers

Product of General Motors
Body by Fisher

New In Style from Radiator to Tail-Light – Offering Scores of Vital Advancements *at No Increase In Price!*

EVEN the impressive array of new features given herewith cannot convey the extent to which the New Series Pontiac Six surpasses all previous attainments in the field of low-priced sixes. After enjoying a spectacularly successful career, Pontiac Six now bids for even greater success with a large roster in every way.

Read This Partial List of Added Features

NEW FISHER BODIES—New lines, new Duco colors, new double beading, more elegant finish, new hood and cowl.

NEW FENDERS—New headlamps and cowl view lamps of matched design with 3 larger and smooth finish.

NEW FOUR-WHEEL BRAKES—Equal to 4-wheel sets, easily operated.

NEW GMR CYLINDER HEAD—Developed by the General Motors Research Laboratories, gives detonation and economical running with any type of fuel.

NEW FUEL PUMP—with gasoline filter. Replaces conventional vacuum tank.

NEW CRANKCASE VENTILATION—eliminates crankcase condensation.

NEW CARBURETOR—with accelerating pump, internal economizer and ventilated choke.

NEW MANIFOLDS AND MUFFLER—for more efficient fueling and exhaust gas scavenging.

NEW AND GREATER POWER—achieved by the foregoing engine improvements.

NEW CROSS-FLOW RADIATOR—has stationary new type cooling system. Virtually eliminates water vapor and alcohol losses. New Indian Head emblem.

NEW THERMOSTAT—automatically assures proper temperature of cooling system water.

NEW WATER PUMP—balanced impeller type with self-feeding bushing.

NEW INSTRUMENT PANEL—oil cluster type. Indirectly lighted.

NEW COINCIDENTAL LOCK—on instrument panel. Turning ignition key also locks transmission.

NEW DASH GASOLINE GAUGE—on instrument panel. Liquid level indicator.

NEW STOP-LIGHT—Tail-light unit.

NEW CLUTCH—single dry plate type. Softer, smoother, more positive in action.

NEW STEERING GEAR—for exceptionally easy steering. New steering wheel.

NEW FRAME—stronger, deeper. Adapted for Lovejoy Shock Absorbers. Includes new tire carrier.

NEW AXLES—front and rear. One inch greater road clearance.

NEW WHEELS—larger and more massive in appearance. New hub flanges.

Emphasizing the importance of this announcement are two entirely new and additional body types, the Four-door Sedan and the Sport Landau Sedan, a close-coupled, swagger creation, exemplifying the highest art of Fisher closed body craftsmanship. Come in and see this history-making line of Sixes, available in six body types.

2-DOOR SEDAN
$745
(At Factory)

COUPE
SPORT ROADSTER
SPORT CABRIOLET
4-DOOR SEDAN
SPORT LANDAU SEDAN

OAKLAND MOTOR CAR COMPANY, 1777 Broadway, New York

MANHATTAN | BRONX | RICHMOND | BROOKLYN

QUEENS | WESTCHESTER

HI-JINKS AT LONDON SCHOOL

The big and little of it. Cyclists do their stuff, straws, toppers and all, in recent London schoolboy hi-jinks.

PACKARD EIGHT PRICES REDUCED

PACKARD'S own Custom Eight cars are today reduced in price as follows:

MODEL	OLD PRICE	NEW PRICE	REDUCTION
Seven Pass. Sedan Limousine	$5250	$4550	$700
Seven Passenger Sedan	5150	4450	700
Two Pass. Convertible Coupe	4950	4250	700
Two Passenger Coupe	4800	4150	650
Five Passenger Club Sedan	4950	4450	500
Four Passenger Coupe	4950	4450	500

The Packard factories are busy to capacity—busier than ever before in nearly thirty years of fine motor car building and at a season of the year when activity is least expected. It is but good business for Packard to share its prosperity with those who buy its products. Therefore the new prices.

There has been no change in quality. Each car is identical with those Packard has been building. The new prices continue to include complete custom equipment and unlimited paint and upholstery options costing hundreds of dollars extra on many other cars. This, together with today's price reductions, gives the Packard Eight an important first cost advantage.

The lower prices make it possible for many additional thousands to step up to the possession of America's finest and most modern car.

We shall be pleased to put this great car into your hands for a demonstration and in return ask only the privilege of telling you what your present car is worth in part payment. Any Packard may be purchased on our payment plan.

(Prices do not include freight and Government tax)

PACKARD

ASK THE MAN WHO OWNS ONE

PACKARD MOTOR CAR CO. OF N. Y.

BROOKLYN BRANCH
Packard Building—Atlantic and Clason Avenues

General Motors Presents
the Reward of a Great Year's Business

Beautiful New
La Salle Family Sedan
with the wonderful Cadillac-
La Salle heavy-duty eight-
cylinder engine

$2350
5-Passenger Sedan, 125-inch wheelbase

$2575
5-Passenger Sedan, 134-inch wheelbase

LA SALLE *follows* CADILLAC
In Lower Prices on the Entire La Salle Line

By marketing more than 15,000 cars in nine months LaSalle has reached the goal set for it as a full year's achievement when Cadillac created this beautiful companion car.

The Cadillac company planned, by giving beauty and value in excess of all previous standards, to win for the eight-cylinder La Salle a market larger than so fine a car had ever been able to command before.

The purpose in this was precisely the same as the policy which has always governed Cadillac and recently resulted in new and lower Cadillac prices—to command by lavish value-giving a demand so great for the La Salle Eight that economies and efficiencies would automatically ensue which would make it possible to *lower prices on the entire La Salle line*.

Enjoy the Prestige and Satisfaction of a La Salle Now

Only a small outlay is required. Appraisal value of your present car acceptable on cash. The balance payable in terms to suit your convenience.

Fifteen thousand loyal and completely contented La Salle owners make that possible today and Cadillac presents La Salle for 1928 as a quality offering so remarkable that it is not even remotely approached by any other car in the world today.

Coincident with this, La Salle offers new types which constitute it outstandingly the American family car of its class—superbly powered by the great Cadillac-La Salle heavy-duty engine and able to out-perform any car in its class or any car now before the public except Cadillac.

A Lower Price on Every Model of La Salle's Complete Line

Two-Passenger Roadster ... $2485	Five-Passenger Sedan ... $2495		
Four-Passenger Phaeton ... $2485	Five-Passenger Town Sedan ... $2495		
Sport Phaeton ... $2975	Five-Passenger Imperial ... $2575		
Two-Passenger Coupe ... $2450	Seven-Passenger Sedan ... $2775		
Two-Passenger Convertible Coupe ... $2550	Seven-Passenger Imperial ... $2875		
Four-Passenger Victoria ... $2550	*All prices f. o. b. Detroit*		

Every La Salle is complete with all modern equipment, much of which is usually obtainable on other cars only at extra cost, and including at the new lower prices, Winter Front, Lovejoy Shock Absorbers, Exclusively Designed La Salle Bumpers, Cowl Lamps, Cord Band, Ventilators, Windshield Wiper, Cigar Lighter, Rear Vision Mirror, etc.

New Additions to La Salle Line

Two-Passenger Business Coupe ... $2350	Five-Passenger Coupe ... $2625
Five-Passenger Family Sedan ... $2350	Five-Passenger Cabriolet Sedan ... $2675
Seven-Passenger Family Sedan ... $2575	

Also Available In Special Fleetwood Custom Built Models

LA SALLE
Companion Car to Cadillac

UPPERCU CADILLAC CORPORATION
INGLIS M. UPPERCU, *President* ARTHUR T. RANDALL, *Manager*
743 Atlantic Avenue, Brooklyn   Telephone Nevins 1539

NEW YORK   NEWARK   NEW ROCHELLE   BRONX   ROCKAWAY   POUGHKEEPSIE   GREAT NECK   YONKERS   WHITE PLAINS

An Important Announcement

by GENERAL MOTORS
concerning the
New CHEVROLET

CHEVROLET GENERAL MOTORS
CONSTANT PROGRESS — BETTER MOTOR CARS

By following a policy of progress in the development of low-priced transportation, General Motors has given the public an ever-increasing measure of—

Modern features—

General Motors, through its Chevrolet division, was a pioneer in giving the public those modern features of design which today are the basis of luxurious low-priced transportation. Prominent among these are: Selective 3-speed transmission; semi-elliptic springs parallel to the frame; electric starting, lighting and ignition; Dash knobs for either water pump cooling system, pressure pump lubrication, and the vacuum fuel system.

Modern performance—

General Motors has always believed that the public is entitled to the advantages inherent in advanced engineering practice. This policy, carried out by the Chevrolet division, has enabled owners of even low-priced cars to enjoy, year after year, the latest developments affecting power, acceleration, smoothness, handling ease, stamina and economy.

Riding comfort—

Realizing that the public's enjoyment of individual transportation is largely dependent on comfort, General Motors has devoted years of research to the basic comfort factors. Numerous discoveries have been made in chassis springs, spring suspension, spring steels, seat cushioning, weight distribution, and body balance. And though Chevrolet prices have been given minimum delay to the purchasers of low-priced automobiles.

Distinctive style—

Early in the development of motor cars, appearance and the pride of possession became important factors in owner satisfaction. Through its long connection with the Fisher Body Corporation, General Motors has consistently introduced finer and more beautiful automobiles. And all the basic advancements in styling have been available to Chevrolet as rapidly as they were created.

As a result of its policy of progress, General Motors through its automobile divisions, has always given the public without delay the benefits of advanced engineering development.

In the short time that has passed since its presentation, the new Chevrolet has become a subject of comment and discussion in homes all over the land.

One of the expressions most frequently heard is: "How can Chevrolet build so fine an automobile and sell it at such low prices?"

To this perfectly natural question there is a perfectly simple answer. The new Chevrolet was produced to sell at its present prices only because the Chevrolet Motor Company used with telling effect the many distinct advantages it enjoys as a division of General Motors.

The style, comfort, performance and quality of the new Chevrolet represent more than the efforts of a single organization engaged in building a single type of car. It embodies in full measure the experience gained by *all* General Motors divisions in the development of *all* General Motors cars.

Its prices are based on the many economies of General Motors' tremendous purchasing power and diversified automotive manufacture.

And its modern design reflects the General Motors policy of constant progress—by which the benefits of engineering advancement are given to the public at the earliest possible opportunity.

General Motors acquires its basic materials in tremendous volume—steel by the hundreds of thousands of tons—wire by the tens of thousands of miles—upholsteries by the acres—nuts, bolts and washers by the millions. This makes it possible to command the most favorable prices from sources of supply whose specialized experience enables them to produce the finest materials for a given purpose. Millions of dollars are saved each year; and millions are passed on to the public in the form of finer, more modern, more desirable automobiles at lower prices.

Through its Fisher Body division, General Motors is the world's largest builder of automobile bodies, with 44 body plants in various parts of the country. It operates its own lumber mills and owns thousands of acres of hardwood forests. It makes its own glass in the world's largest plate glass factory. It manufactures all of its own body hardware. Its craftsmanship is internationally renowned, and its style influence is felt throughout the industry.

As a result, General Motors cars in every price class have Fisher bodies of acknowledged distinction and sound construction. And nowhere is this more vividly revealed than in the new Chevrolet.

General Motors manufactures, both for itself and the industry at large, an almost limitless number of specialized automotive products. Starting, lighting and ignition equipment—ball bearings—steering wheels—roller bearings—warning horns—wheels—rims—spark plugs—air cleaners—oil filters—radiators—all call for specialized engineering and production skill.

General Motors makes them all—and purchasers of motor cars the world over benefit accordingly.

The engineering staffs responsible for all General Motors cars are continually striving for the new and better thing. At their disposal are the General Motors Research Laboratories and the General Motors Proving Ground. Every engineering advancement and scientific discovery resulting from this combined effort is immediately available for use on all General Motors automobiles from Cadillac to Chevrolet.

This is an advantage of untold value. Years ago it enabled Chevrolet to pioneer into the low-priced field those quality features on which the present conception of a low-priced quality car is based. Today, it makes possible numerous new improvements in the Bigger and Better Chevrolet.

In developing the Chevrolet cars of the past, General Motors has made a vital contribution to the happiness and welfare of the nation.

It created an entirely new idea of what the buyer of a low-priced automobile could expect. It made luxurious transportation the pleasure of the many, rather than the privilege of the few. It has made *progress* an automotive watchword.

Now, in the new Chevrolet, that policy of progress finds still further expression. More modern features have been provided. More distinguished style has been created. Finer performance has been attained. Greater comfort has been provided. And wider public service has been rendered.

GENERAL MOTORS
"A car for every purse and purpose"

CHEVROLET · PONTIAC · OLDSMOBILE · OAKLAND · BUICK · LaSALLE · CADILLAC
All with Body by Fisher
GENERAL MOTORS TRUCKS · YELLOW CABS and COACHES
FRIGIDAIRE—The Electric Refrigerator   DELCO-LIGHT Electric Plants

G. M. A. C.—The Payment Plan for General Motors Products

Everywhere
Tremendous Enthusiasm for the
New Chevrolet

Offering marvelous new Fisher bodies, styled with all the artistry of world-famous master designers... embodying scores of new engineering advancements... and providing a new and thrilling type of performance—the Bigger and Better Chevrolet is everywhere exciting wildfire enthusiasm.

Study this great new car and you will discover countless new features that typify the progressiveness which the public has come to expect from Chevrolet: a stronger, sturdier frame, with a 107 inch wheelbase—4 inches longer than before; non-locking 4-wheel brakes; semi-elliptic shock absorber springs, 84% of wheelbase; ball bearing worm and gear steering mechanism; alloy "Invar strut" pistons; completely enclosed instrument panel indirectly lighted; deeper radiator with automatic thermostat control; larger balloon tires 30" x 4.50"; and many others in addition to the world-famous features which Chevrolet pioneered into the low-price field.

In spite of all the engineering advancements it has made from year to year, the Chevrolet Motor Company has never before presented an achievement so sensational as the Bigger and Better Chevrolet. Come in and see it today!

PRICES REDUCED!

The COACH **$585**

The Roadster	$495
The Touring	$495
The Coupe	$595
The 4-Door Sedan	$675
The Sport Cabriolet	$665
The Imperial Landau	$715
Light Delivery	$375
Utility Truck	$490

Bigger and Better — 4 Wheel Brakes

[Dealer listings for Brooklyn and Queens]

QUALITY AT LOW COST

DETONOX GASOLINE

YEAR 'ROUND USERS of DETONOX often remark about its *uniform* quality.

They say the DETONOX they bought last night, and last year —the DETONOX they bought in New York, and in Minneapolis —was *identical* in performance.

Always the Same Goodness!

DETONOX quality is uniform, because exactingly controlled. Each "run" of this gasoline thru Pure Oil's own modern refineries is repeatedly tested for quality. And no DETONOX is shipped until expert chemists certify that it conforms to rigid specifications.

Is it any wonder that in DETONOX you get *unvarying* high quality?

Better 7 Ways
1. Non-detonating
2. Instant starting
3. Lightning pick-up
4. Giant power
5. Better mileage
6. Clean, less carbon
7. Safe for you and your motor

Look for the Red Gasoline in the "Pure Oil Blue" Pumps

Product of THE PURE OIL COMPANY, U.S.A.

KISSEL

Announcing

THE APPOINTMENT OF
DALE & de RHAM MOTORS, INC.

KISSEL DISTRIBUTOR
463 PARK AVE. (RITZ TOWER)
PLAZA 3275

And a Preliminary Display of the

New Smaller Eight

which will be shown at the Coming New York Show.

Coincident with the appointment of Dale & de Rham, Inc., new Kissel distributor, comes the important announcement of the Smaller Eight—a new eight of Kissel quality, smaller in size, smaller in price—$1895. This beautiful new car is now on display at Dale & de Rham's showrooms, prior to its introduction at the New York Show.

Both Kissel and Dale & de Rham cordially invite you to view and inspect this car—and also the New Smaller Six at $1495, the De Luxe Model 90 and several new body styles.

Also learn about the 1928 Sensation—the new Kissel White Eagle, 115 Horsepower Motor—*100 miles an hour.*

KISSEL MOTOR CAR CO.
Hartford, Wis.

TO-NIGHT

10.30 P.M. to 11.30 P.M.
Hear the Most Unusual Radio Program Ever Broadcast

DODGE BROTHERS
VICTORY SIX
RADIO HOUR

WILL ROGERS — Master of ceremonies throughout the program, speaking from his home in Beverly Hills, California.

AL JOLSON — In a series of characteristic Jolson songs, from his hotel in New Orleans.

PAUL WHITEMAN — With the internationally famous Whiteman band in New York. Mr. Whiteman himself will announce each selection.

FRED AND DOROTHY STONE — In a typical Stone program — Fred talking, Dorothy singing — in Chicago.

Tonight all America will become one vast radio audience.

Maximum facilities of the National Broadcasting Company and the American Telephone and Telegraph Company will be utilized. 33 broadcasting stations, 3 transcontinental telephone circuits with 12,000 miles of wire, scores of engineers and more than 200 station operators will cooperate in the most extensive broadcasting hook-up ever attempted.

Will Rogers will act as master of ceremonies throughout the hour. His inimitable quips will be carried from Hollywood, California, to station WEAF, New York, and from there will be retransmitted to the world.

From New Orleans, Al Jolson will sing, as only Jolson can. Fred and Dorothy Stone will entertain in Chicago, while from New York the music of Paul Whiteman's band, Paul Whiteman himself wielding the baton, will be broadcast to the four corners of the nation.

President E. G. Wilmer will announce Dodge Brothers epochal new car — The Victory Six — the most spectacular engineering triumph of the decade, and the only car of its kind in the world.

Remember the hour, 10:30 to 11:30 Eastern Standard Time — tonight.

WEAF

COMPLETE RACING RESULTS
IN THE EAGLE
SPORT EDITIONS EVERY DAY

WATCH FOR THEM

BRAKES

COME to Brooklyn's largest brake service organization when you need brake service. 5 stations. Prompt service. Courteous attention. Inspection—adjustment—equalization and relining of all types of brakes by experts.

OTHER STATIONS
Brooklyn
1753 Bedford Ave.
1064 Bedford Ave.

Bay Ridge
6262 Sixth Ave.

Long Island City
3036 Northern Blvd.

TILDEN'S
1030 ATLANTIC AVE.
(Between Classon and Grand Aves.)

WE WILL EXHIBIT AT THE BROOKLYN AUTO SHOW

28th Annual NATIONAL AUTO SHOW

Opens SATURDAY, 2 P.M. Daily Thereafter (Except Sunday) 10 A.M. to 10:30 P.M.

JAN. 7 to 14

The Newest in Cars, Accessories and Light Trucks—Shop Equipment Section Open to Public After 5 P.M. Daily

Two entrances, Park Ave. and Lexington Ave.

GRAND CENTRAL PALACE

Adm. 75¢

CADILLAC
*presents a new V·8 * V·12 * and La Salle*
AT DECIDED REDUCTIONS IN PRICE

Re-styled completely in a mode quite obviously their own ... endowed with performance more brilliant and alluring from every standpoint ... and enriched throughout in those fundamental qualities which set Cadillac-built cars apart from all others in the world—a new Cadillac V-8, a new Cadillac V-12 and a dashing new La Salle now make their bow to the American public. ° ° ° And they come not only as the finest cars that Cadillac has ever built, but as by far the greatest values—for every model in all three lines is offered at prices decidedly reduced. ° ° ° At no time in its history has Cadillac made a more significant step toward its goal of perfection than in these three distinguished cars. , In everything that they are and do, they represent definite advancement over their distinguished predecessors. ° ° ° They are infinitely more beautiful—not only in the sweep of their lines, but in the contour of radiator, fenders and mouldings. And their beauty is individual, lending a distinction apart from all other cars on the highway—as you would expect of a Cadillac or La Salle. ° ° ° Every performance factor is decisively improved and refined. The engines are more alert and dynamic—not only in speed, but in response to the throttle as well. The over-all quietness is even more pronounced, actually approaching complete silence of operation. Comfort is the nearest approach to the ideal the motor car has ever known. And there is a quick, certain and unlabored response to brakes and steering wheel that removes from driving the last vestige of effort. ° ° ° And Cadillac is especially proud to provide, alone of all the cars in its price field, that vital health and comfort feature—Fisher No-Draft Ventilation, individually-controlled. It may be said definitely that, without this feature, no car can give you the satisfaction you now have a right to expect in a fine automobile. ° ° ° These three unusual cars are now on display at Cadillac-La Salle showrooms everywhere, and Cadillac joins with its dealers in extending a most cordial invitation to see and inspect them at any time.

° ° °

As previously announced, the 1933 production of the Cadillac V-16 will be limited to four hundred cars, custom built to individual order—each car to be serially numbered and inscribed with the owner's name. The coachwork will be designed and executed by Fleetwood and will be mounted on an improved 16-cylinder chassis of even greater performance and mechanical excellence. For those who seek the acme in personalized transportation, the Cadillac V-16 comprises its highest interpretation.

° ° La Salle list prices now from $2245, Cadillac list prices now from $2695, f. o. b. Detroit ° °

CADILLAC MOTOR CAR COMPANY

"I'm a conservative driver–but I want an oil that can hit 100!"

CONSERVATIVE speeds may suit you to a T. You may never want to drive faster than 40-miles-an-hour...

But whether you speed or not, the oil you use should be the 100-miles-an-hour oil —GULF SUPREME! And here's why...

The oil that can win out under the terrific punishment of 100-miles-an-hour is a *better, safer oil*. It's good at high speeds— and *doubly good at lower speeds*!

It gives you more protection against wear. It gives you extra richness. Extra stamina. Extra lubrication!

And here's proof that Gulf Supreme can take super-punishment... It successfully lubricated a motor running

at almost twice the heat of a speeding engine... for 14 solid hours!

And it amazed race-drivers at the Indianapolis Speedway by out-performing special "racing oils"! It lubricated a thundering Duesenberg racer—under Official AAA supervision—at speeds almost as high as two miles a minute. An average speed for the one-hour, non-stop run of better than 100-miles-an-hour!

Drive into any Gulf station now. Drain worn oil. Refill with Gulf Supreme. There's a grade for every climate. Its ability to take punishment means money in your pocket—and longer life for your motor!

WARNING!
...OIL that isn't good at high speeds, isn't good enough at ANY speed!

GULF SUPREME MOTOR OIL
"The 100-Mile-An-Hour Oil"

AND WHEN YOU BUY GASOLINE... GET THAT GOOD GULF ...it's fresh!

Tel. MIDWOOD 9726
BROOKLYN AUTO RENTAL SERVICE

New Cadillac Limousines for all occasions, by the hour, day, week or month

ARTHUR KINSLEY   1328 E. 37th St.

Excellent Reasons for Attending the Motor Car Show

CADILLAC V-8
and
LA SALLE V-8

manufactured and produced in the same world-famous precision laboratories of the Cadillac Motor Car Company which designed, built and presented for the first time at the 30th Annual Automobile Show (New York) the new sixteen-cylinder Cadillac to be shown in this city at a later date.

—

Three cars built in the same highly specialized shops and by the same experts in precision manufacture... All three drawing from the same rich sources—General Motors, Cadillac, Fisher and Fleetwood... All three profiting by 27 years of exclusive devotion to the creation of the finest possible motor cars. Cadillac Motor Car Company, Division of General Motors.

UPPERCU CADILLAC CORPORATION

INGLIS M. UPPERCU, *President*   ARTHUR E. RANDALL, *Vice-President*

749 ATLANTIC AVENUE   8703 FOURTH AVENUE

NEVINS 2500   OPEN EVENINGS   SHORE ROAD 7000

Acknowledgements

This book was compiled with a little help from my friends, and with copyediting by Lauren Moccio. If you would like to join in the fun or just keep in touch with us—or to find some of your old friends from Brooklyn—social media is the way to do it. Please visit www.parkslopian.com or www.brooklynpast.com for links to more memories and pictures of a Brooklyn long past, and the iconic things of yesteryear. I, for one, would love to hear from more of my old friends; how about you?

Brooklyn the series: In which neighborhood in Brooklyn did you grow up? Do you yearn for the good old days? This series of neighborhood books is the closest you'll get to reliving the past—or, if you're not from Brooklyn, experiencing the fun, the excitement and the tragedies, right from the memories of the kids and the parents who lived there during the fascinating middle of the twentieth century. These books walk you through the various neighborhoods in classic 1950s through 1980s Brooklyn, detailing the iconic things of our time. From the doctors who delivered us to the schools from which we graduated, from the playgrounds and parks in which we played to the street games we made up by ourselves, from all the great toys we had that have since been replaced by sharper technology, our first bikes to our first cars—we remember it all. Take a trip down memory lane with these coffee table books, written in an enjoyable, accessible, social media style. Revisit all the best places we ate; remember all the silly slang and the nonsensical stuff we used to say. As with every passing generation, we cling to the things that defined our youth. However, we who grew up in the fifties through the eighties experienced some of the most timeless pop culture in history, and this extraordinary series will allow you to share that with your children using a language they understand: social media.

This wonderful series of coffee table books contrasts bonding over modern social media with longing for the past. If aliens came down to Earth sometime in the future and found these books, they would act as a time capsule—a treasure trove of memories of mid-twentieth century pop culture, and a demonstration of twenty-first century netspeak and social media usage to boot. Use these books to share your memories with friends who grew up elsewhere, and as you pore over the pages, find all the similarities and differences in the toys you owned versus the mom and pop businesses that you frequented. Designed as coffee table books, they can be picked up and put down at your leisure; they do not need to be read cover to cover. If you own these books and display them in your living room, your friends will want to visit you more often!

Check our site for new books coming soon, Including **Brooklyn Matchbooks, Brooklyn Postcards, Brooklyn Auto Ads, Brooklyn Business Cards, Brooklyn Envelopes** and these great Brooklyn neighborhoods Sheepshead Bay, Canarsie, Mill Basin, Boro Park, Williamsburgh, South Brooklyn, Carroll Gardens, Gravesend, Marine Park, Coney Island, and many more..

Brooklyn The series: The Parkslopian , Growing up in Brooklyn, Growing up in Bay Ridge ..

Made In Brooklyn : Vol: 1. Featuring over 400 vintage ads, matchbooks, and other types of Brooklyn Ephemeral Dating back to the late 1800's

Brooklyn Vintage Ads: Vol: 2. Featuring over 400 vintage ads, matchbooks, and other types of Brooklyn Ephemeral Dating back to the late 1800's

Brooklyn has always been the most populous of New York City's five boroughs, with a steady population of 2.5 million since 1950s. It is said that one in seven Americans can trace their roots to Brooklyn. Today, if Brooklyn were an independent city, it would rank as the fourth most populous city in the United States. Actually, Brooklyn was once a city! All of these people can find a piece of their heritage within the pages of these books. This niche easily expands into the iconic gems of the past that were shared by everyone in the country, and the second book in the series, Iconic Things of Our Time, will focus on the images of the time period that can be shared by all Americans. This series allows you to choose your focus: take a broad view of the country in the 50s through the 80s, or narrow your scope to individual neighborhoods in Brooklyn—then, compare and contrast what we shared and what made us unique. Travel back in time in the comfort of your own living room, with your loved ones, with your children, with your friends-remember all these iconic things, one page at a time.

Links @ **http://www.BROOKLYNPAST.COM**

Get your very own Brooklyn T-Shirt
All Neighborhoods are represented here in this one of a kind Brooklyn shape Design of the Brooklyn subway system, with vintage style lettering spelling out BROOKLYN and iconic pictures placed inside the letters....

Style 2: T-Shirt. All Neighborhoods are represented here in this one of a kind Brooklyn shape Design.. Brooklyn shaped design with all 60 neighborhoods listed in this cloud style design…

Show your Brooklyn pride with these two great looking shirts available at Brooklynpast.com

COMING SOON GET THESE 2 GREAT DESIGNS.. 8 X 10 PHOTO TO FRAME AND KEEP BROOKLYN IN YOUR HEART FOREVER..
Brooklynpast.com

www.ingramcontent.com/pod-product-compliance
Lightning Source LLC
Chambersburg PA
CBHW071629220526
45469CB00002B/538